RECENT ADVANCES IN NUMERICAL ANALYSIS

Publication No. 41
of the Mathematics Research Center
The University of Wisconsin–Madison

ACADEMIC PRESS RAPID MANUSCRIPT REPRODUCTION

RECENT ADVANCES IN NUMERICAL ANALYSIS

Edited by

CARL DE BOOR
Mathematics Research Center
University of Wisconsin
Madison, Wisconsin

GENE H. GOLUB
Computer Science Department
Stanford University
Stanford, California

Proceedings of A Symposium
Conducted by the Mathematics Research Center
The University of Wisconsin – Madison
May 22– 24, 1978

ACADEMIC PRESS New York San Francisco London 1978
A Subsidiary of Harcourt Brace Jovanovich, Publishers

ACADEMIC PRESS, INC.
111 Fifth Avenue, New York, New York 10003

United Kingdom Edition published by
ACADEMIC PRESS, INC. (LONDON) LTD.
24/28 Oval Road, London NW1 7DX

Library of Congress Cataloging in Publication Data

Main entry under title:

Recent advances in numerical analysis.

Includes index.
1. Numerical analysis--Congresses. I. de Boor, Carl. II. Golub, Gene Howard, Date III. Wisconsin. University--Madison. Mathematics Research Center.
QA297.R39 519.4 78-25621
ISBN 0-12-208360-1

PRINTED IN THE UNITED STATES OF AMERICA

78 79 80 81 82 9 8 7 6 5 4 3 2 1

Contents

Contributors

Numbers in parentheses indicate the pages on which the authors' contributions begin.

Germund Dahlquist (1), Department of Numerical Analysis, Royal Institute of Technology, S-10044 Stockholm, Sweden

Todd Dupont (31), Department of Mathematics, University of Chicago, Chicago, Illinois 60637

Walter Gautschi (45), Department of Computer Sciences, Purdue University, West Lafayette, Indiana 47907

Herbert B. Keller (73), Applied Mathematics 101-50, California Institute of Technology, Pasadena, California 91125

Heinz-Otto Kreiss (95), Applied Mathematics, California Institute of Technology, Pasadena, California 91125

Peter D. Lax (107), Courant Institute of Mathematical Sciences, New York University, New York 10012

Robert E. Lynch (143), Division of Mathematical Sciences, Purdue University, West Lafayette, Indiana 47907

J. A. Nitsche (119), Institut für Angewandte Mathematik, Albert-Ludwigs-Universität, Hermann-Herder-Str. 10, 7800 Freiburg i. Br., Federal Republic of Germany

John R. Rice (143), Division of Mathematical Sciences, Purdue University, West Lafayette, Indiana 47907

Ridgway Scott (31), Mathematics Department, The University of Michigan, Ann Arbor, Michigan 48104

L. F. Shampine (177), Applied Mathematics Research Department, Sandia Laboratories, Albuquerque, New Mexico 87185

G. W. Stewart (193), Department of Computer Science, University of Maryland, College Park, Maryland 20742

H. F. Weinberger (207), School of Mathematics, University of Minnesota, 206 Church Street, Minneapolis, Minnesota 55455

J. H. Wilkinson (231), Department of Computer Science, Stanford University, Stanford, California 94305

Preface

This book contains the Proceedings of the Symposium on Recent Advances in Numerical Analysis held in Madison, Wisconsin, May 22–24, 1978, under the auspices of the Mathematics Research Center, University of Wisconsin–Madison, and with financial support from the United States Army under Contract No. DAAG29-75-C-0024. Due to the necessity of meeting the publication schedule, the text of Professor W. Kahan's fine lecture has not been included in this volume.

The symposium was dedicated to Professor J. Barkley Rosser, Professor of Mathematics and Computer Science, University of Wisconsin–Madison, since 1963, and Director, Mathematics Research Center, 1963 to 1973, on the occasion of his retirement.

We thank the authors for their efforts to meet our manuscript deadline, and thank Dorothy M. Bowar who put the volume together and compiled the index. We are especially grateful to Gladys Moran, our experienced symposium secretary, who so ably handled the many organizational details.

Carl de Boor and Gene H. Golub

J. Barkley Rosser

Dedicated to Professor J. Barkley Rosser

Positive Functions and some Applications to Stability Questions for Numerical Methods

Germund Dahlquist

1. INTRODUCTION

In this paper, an analytic function f is called positive, or in short

$$f \in P,$$

if the following conditions are satisfied:

A. $f(z)$ is regular for $\operatorname{Re} z > 0$,

B. $\operatorname{Re} f(z) > 0$ for $\operatorname{Re} z > 0$.

In other words, f maps $\mathbb{C}^+ = \{z : \operatorname{Re} z > 0\}$ into $\mathbb{C}^+$. Note that $\mathbb{C}^+$ denotes the open right half-plane.

We shall also talk about the slightly larger class P', where condition B is replaced by

B'. $\operatorname{Re} f(z) \geq 0$ for $\operatorname{Re} z > 0$.

By the minimum principle, the infimum of the harmonic function $\operatorname{Re} f(z)$ cannot be reached at an interior point of $\mathbb{C}^+$, unless $\operatorname{Re} f(z) \equiv 0$. It follows that $P' \setminus P$ contains only those functions for which $\operatorname{Re} f(z) \equiv 0$. Then, by the Cauchy-Riemann differential equations, $\operatorname{Im} f(z) \equiv \text{const}$, i.e.

$$P' \setminus P = \{f : f(z) = \text{imaginary constant}\}.$$

In this paper the term "imaginary" includes 0, and whenever it makes sense, also ∞.

If a sequence of functions $\{f_n\}$, $f_n \in P'$, converges to a limit function f that is analytic in $\mathbb{C}^+$, then $\operatorname{Re} f(z) \geq 0$ for $z \in \mathbb{C}^+$, hence $f \in P'$.

ISBN 0-12-208360-1

Note that a positive function is allowed to have singularities on the imaginary axis (including 0 and ∞). For example, the function z^{-1} is positive with a pole at 0, the function z is positive with a pole at ∞. The functions $\tanh z$ and $\log(1+z)$ are positive with essential singularities at ∞.

For convenience we shall usually use sloppy expressions like $z^{-1} \in P$, instead of the more correct expression $f \in P$, where f is defined by $f(z) = z^{-1}$.

The class P has some useful properties to bear in mind that follows directly from the definition, e.g. if $f \in P$, $g \in P$, then,

1) $\alpha f + \beta g \in P$, $\alpha > 0$, $\beta \geq 0$,
2) $g(f(z)) \in P$, in particular,
2a) $1/f(z) \in P$
2b) $\alpha f(z) + \gamma \in P$
2c) $g(\alpha z + \gamma) \in P$

(2b, 2c: α, γ constants, $\alpha > 0$, $\operatorname{Re}\gamma \geq 0$)

Note however that in general $f \cdot g \notin P$, e.g. $z^2 \notin P$, but $(f \cdot g)^{1/2} \in P$ because

$$|\arg(f(z)\ g(z))^{1/2}| \leq \frac{1}{2}\left(|\arg f(z)| + |\arg g(z)|\right) < \frac{\pi}{2}\ .$$

The term "positive" is borrowed from electrical engineering, where this class of functions and related classes play an important role in the synthesis of passive networks after the pioneering work of Brune [6] in 1931. We quote from Šiljak [33]: "It not only laid the basis for all of the realization theory of electrical networks, but was subsequently utilized in areas as diverse as absolute stability and hyperstability, optimality and sensitivity of dynamical systems." See also Guillemin [24, p. 409], Belevitch [4 , Ch. 5].

Related classes of functions have been studied in several branches of mathematics. Functions mapping the upper half-plane into itself, sometimes called Pick functions [16], are important in the moment problem [32], or in spectral theory for operators in Hilbert space, see e.g. [1 , Ch. 6]. There is also an extensive literature about functions, which are bounded [31] in the unit disk. We regard "the class P" as a convenient normal form for such problems. It is a matter of taste whether, in a particular application, one should transform a problem concerning a related class to a "class P" problem, or one should formulate "P class" properties in terms of the function class of the original statement of the problem.

The purpose of this article is to give evidence for the following statements.

a) Many stability conditions for numerical methods can be expressed in the form that some associated function is positive (Chapters 2 and 4).

b) Positive functions have many pleasant, well-known properties, which are simple to use in investigations what stability and accuracy properties can be possessed by a numerical method of a certain type (Chapters 3-6). The emphasis of this paper is on the applicability of these properties to numerical analysis. Except for the expansion in Section 4.3, we shall derive no new facts about positive functions. Some proofs may be new.

Most often, though not always, the positive function is real on the positive real axis, e.g. a rational function with real coefficients. The function is then called positive real. Many calculations and derivations are simpler in this particular case.

2. FORMULATION OF SOME STABILITY CONDITIONS FOR NUMERICAL METHODS IN TERMS OF POSITIVE FUNCTIONS

2.1. LINEAR MULTISTEP METHODS

Consider a linear multistep method

$$\sum_{j=0}^{k} \alpha_j y_{n+j} = h \sum_{j=0}^{k} \beta_j f(t_{n+j}, y_{n+j}) , \tag{2.1}$$

for the numerical treatment of the differential system, $y' = f(t,y)$. Let the generating polynomials be

$$\rho(\zeta) = \sum_{j=0}^{k} \alpha_j \zeta^j, \qquad \sigma(\zeta) = \sum_{j=0}^{k} \beta_j \zeta^j . \tag{2.2}$$

For the linear test-problem, $y' = \lambda y$, we obtain a linear difference equation with the characteristic equation

$$\rho(\zeta) - \lambda h \sigma(\zeta) = 0 ,$$

or if we put $\lambda h = q$,

$$q = \rho(\zeta)/\sigma(\zeta) . \tag{2.3}$$

A-stability [9] means that we require that all solutions of the difference equation should be bounded when $n \to \infty$, whenever $\operatorname{Re} q \leq 0$. This is a natural, though not necessary, requirement for stiff problems. This is equivalent to saying that if $|\zeta| > 1$ then q must have a positive real part. Then, by (2.3),

$$\mathrm{Re}\ \rho(\zeta)/\sigma(\zeta) > 0, \qquad \text{when} \quad |\zeta| > 1\ . \tag{2.4}$$

We shall here, and in other contexts in this paper, utilize the transformation

$$z = z(\zeta) = (\zeta + 1)/(\zeta - 1)\ , \tag{2.5}$$

that maps the exterior of the unit disk of the ζ-plane onto $\mathbb{C}^+$ of the z-plane,

$$|\zeta| > 1 \Longleftrightarrow \mathrm{Re}(\zeta + 1)/(\zeta - 1) > 0\ , \qquad \zeta = 1 \Longleftrightarrow z = \infty, \tag{2.6}$$

and vice versa. Note that the mapping is involutary, i.e.,

$$\zeta = (z(\zeta) + 1)/(z(\zeta) - 1), \tag{2.7'}$$

and that $\rho(\zeta)/\sigma(\zeta)$ becomes a rational function $r(z)/s(z)$, which will be called the canonical function of the method, Genin [20],

$$\rho(\zeta)/\sigma(\zeta) \equiv r(z)/s(z)\ . \tag{2.7''}$$

The A-stability condition then reads

$$r(z)/s(z) \in P\ . \tag{2.8'}$$

Since the reciprocal of a function in P also belongs to P (Property 2a of Chapter 1) we can equivalently write the condition in the form,

$$s(z)/r(z) \in P\ . \tag{2.8''}$$

Using a general representation formula for positive functions, see Chapter 5, the author proved [9] that an A-stable linear multistep method cannot be more than second order accurate. See another treatment in Section 4.2. One therefore has to consider less demanding stability requirements. It can be shown, see e.g. [13, Section 5] that the requirement that the region of absolute stability should contain the domain,

$$D(a,b,c,d) = \{q : \mathrm{Re}\ \frac{aq + b}{cq + d} \le 0\}\ , \tag{2.9}$$

can be written

$$\frac{ar + bs}{cr + ds} \in P \tag{2.10}$$

Any circular disk, or its exterior, as well as any half-plane can be written in the form D(a,b,c,d) by an appropriate choice of a, b, c, d. The "spirit" of the notion of stiff stability, Gear [19], can be obtained by formulating that the region of absolute stability should contain the union of two (or perhaps three) domains of this type. The properties of positive functions give hints for a neat formulation of stability and accuracy requirements as a system of inequalities and equations. Genin [20] investigates stiffly stable three-step methods from this point of view in a paper that contains many new ideas.

The notion of $A(\alpha)$-stability, Widlund [36], can be expressed in the form,

$$(r/s)^{1/(2-2\alpha/\pi)} \in P , \qquad 0 < \alpha \leq \pi/2 ,$$

although we have seen no application of this formulation yet. For $\alpha = 0$ one obtains A_0-stability, Cryer [7]. Another formulation of $A(\alpha)$-stability, Bickart and Jury [5] is obtained by considering the sector $|\arg q - \pi| < \alpha$ as the union of a one-parameter family of circles, $|q + \beta| \leq \beta \sin \alpha$, $0 < \beta < \infty$, or equivalently, $\mathrm{Re}(q + \beta + \beta \sin \alpha)/(q + \beta - \beta \sin \alpha) \leq 0$.

It has recently been shown, Dahlquist [10], Liniger and Odeh [30], Dahlquist and Nevanlinna [14], Kreiss [28] that these stability concepts are relevant for the numerical stability of these methods in the application to a large class of non-linear systems. Dahlquist [13], shows this by a new general expansion formula for positive functions, see also Section 4.3 of the present paper.

In most applications, the positive functions are rational functions. One exception is a recent application [12] to linear multistep methods of the form,

$$\sum_{j=0}^{k} \alpha_j y_{n+j} = h^2 \sum_{j=0}^{k} \beta_j f(t_{n+j}, y_{n+j}) , \tag{2.11}$$

for special second order differential systems, $y'' = f(t,y)$. A method is said to be <u>unconditionally stable</u>, if it produces bounded solutions, when it is applied with fixed step size h to any test equation of the form, $y'' = -w^2 y$, $w > 0$. Analogously to the A-stability case for first order differential equations, this leads to the requirement that $r(z)/s(z)$ must not be real and negative for $\mathrm{Re}\, z > 0$, which is equivalent to the statement that

$$(r/s)^{1/2} \in P . \tag{2.12}$$

We shall call $(r/s)^{1/2}$ the <u>canonical function</u> for the method defined by (2.11). Note that, unlike most applications of positive functions, this function is not rational. It is shown in [12], see also Section 4.2, that this is incompatible with more than second-order accuracy.

2.2. RUNGE-KUTTA METHODS

When an s-stage Runge-Kutta method is applied to the linear test equation, $y' = \lambda y$, with fixed step size h, one obtains a recurrence relation of the form,

$$y_{n+1} = \frac{N(q)}{D(q)} y_n , \qquad (q = \lambda h) ,$$

where $N(q)$ and $D(q)$ are polynomials of degree s at most. If the method is explicit, then $D(q)$ is a constant. $N(q)/D(q)$ should be an approximation to $\exp q$, when $|q|$ is small. The region of absolute stability is equal to

$$S = \{q: |N(q)/D(q)| \leq 1\} .$$

By (2.6), this can also be written in the form,

$$S = \left\{ q: \mathrm{Re}\, \frac{D(q) + N(q)}{D(q) - N(q)} \geq 0 \right\} \tag{2.13}$$

Let $q(z)$ be an analytic function that maps $\mathbb{C}^+$ one-to-one onto a set S'. The statement $S \supset S'$ is then equivalent to the statement,

$$\frac{D(q(z)) + N(q(z))}{D(q(z)) - N(q(z))} \in P . \tag{2.14}$$

Jeltsch and O. Nevanlinna [27] have recently used this formulation and general properties of positive functions to show that the region of absolute stability of an explicit, s-stage Runge-Kutta method can contain the disk $\{q: \mathrm{Re}(q + 2R)/q \leq 0\}$ only if $R \leq s$. For $R = s$, this circle, for which the interval $[-2s, 0]$ is a diameter, is the region of absolute stability of the s-stage method, which consists of s Euler steps, each of length h/s.

The theory of positive functions plays, however, no role in two recent, very different, proofs, Gobert [23], Wanner, Hairer and Nørsett [35], of the Ehle conjecture concerning the A-stability of Runge-Kutta methods for which $N(q)/D(q)$ is a Padé approximant of the exponential.

2.3. OTHER APPLICATIONS OF POSITIVE FUNCTIONS

A polynomial,

$$g(z) = \sum_{j=0}^{k} (\gamma'_j + i\gamma''_j) z^j , \qquad (\gamma'_j, \gamma''_j \text{ real}) \tag{2.15}$$

is called a <u>Hurwitz polynomial</u>, if all zeros are located in $\mathbb{C}^-$. Many stability questions can be reduced to the question whether or not a certain polynomial is a Hurwitz polynomial.

There are several connections between Hurwitz polynomials and positive rational functions. First, a rational function $r(z)/s(z) \in P'$, <u>iff</u> $r(z) - qs(z)$ <u>is a Hurwitz polynomial for all</u> $q \in \mathbb{C}^-$. Second, let

$$\bar{g}(z) = \sum_{j=0}^{k} (\gamma'_j - i\gamma''_j)z^j ,$$

$$g_1(z) = (g(z) - \bar{g}(-z))/2 , \qquad (2.16)$$

$$g_2(z) = (g(z) + \bar{g}(-z))/2 .$$

<u>LEMMA</u>. g is a Hurwitz polynomial, iff
 i) g_1 and g_2 have no common divisor,
ii) $g_1/g_2 \in P$. □

This lemma is a formulation of a step used in a proof by Frank [17], see also Wall [34] for the real case. Condition ii) is, by (2.6), equivalent to stating that

$$|g(z)/\bar{g}(-z)| > 1 \quad \text{for} \quad z \in \mathbb{C}^+ ,$$

One can show [34] that this is a characteristic property of Hurwitz polynomials, by regarding $|g(z)|$ as the product of the lengths of the vectors from z to its zeros.

Note that $\varphi = g_1/g_2$ satisfies the relation,

$$\varphi(z) = -\bar{\varphi}(-z) . \qquad (2.17)$$

Such functions are called <u>para-odd</u> in circuit theory. We shall see in Section 3.2 that a characteristic property of para-odd positive functions is that they can be represented by a continued fraction, where certain coefficients are positive, and that this yields a criterion for Hurwitz polynomials, which is an extension of the Routh algorithm to complex polynomials, due to Frank [17].

In Section 4.1 we give an example from a stability analysis of a method proposed for hyperbolic initial-boundary value problems, where it is to be shown that a certain sequence of polynomials, $D_2(z)$, $D_3(z)$, $D_4(z)$, ... have no zeros in $\mathbb{C}^+$. This is achieved by proving by induction that the functions $D_n(z)/D_{n-1}(z) \in P$.

3. SOME PROPERTIES OF POSITIVE FUNCTIONS, WITH APPLICATIONS

3.1. THE BEHAVIOR AT INFINITY AND NEAR IMAGINARY SINGULAR POINTS

Consider the function,

$$f(z) = az^{\alpha}, \qquad \alpha \text{ real, a complex, } a \neq 0 .$$

If we define the argument of a complex number such that $-\pi < \arg w \leq \pi$, then $w \in \mathbb{C}^+ \iff |\arg w| < \pi/2$. It follows that

$$az^\alpha \in P' \qquad \text{if} \quad |\arg a| \leq (1 - |\alpha|)\pi/2 \tag{3.1}$$

It follows that $|\alpha| \leq 1$, and that for $\alpha = \pm 1$, a has to be real and positive.

The same reasoning shows that if $f \in P'$ and if

$$f(z) \sim a(z - i\eta)^\alpha \qquad \text{when} \quad z \to i\eta\ , \quad (a \neq 0)\ , \tag{3.2'}$$

or

$$f(z) \sim az^\alpha\ , \qquad \text{when} \quad z \to \infty\ , \quad (a \neq 0)\ , \tag{3.2''}$$

then

$$|\arg a| \leq (1 - |\alpha|)\pi/2\ , \qquad |\alpha| \leq 1\ . \tag{3.2'''}$$

Note in particular, that $a > 0$ for $\alpha = \pm 1$. The following important results are immediate consequences.

THEOREM. A pole on the imaginary axis (including 0 and ∞) of a positive function must be simple and have a real and positive residue. A zero on the imaginary axis must be simple.

PROPOSITION. The modulus of the difference of the degrees of the numerator and the denominator of a positive rational function cannot exceed unity.

Also note that, since poles are the only singularities of a rational (or a meromorphic) function, the regularity condition A (see Chapter 1) for such functions follows from condition B, since the argument variation of $f(z)$ on a small contour around an interior pole would be at least 2π, which is incompatible with $|\arg f(z)| < \pi/2$ for $z \in \mathbb{C}^+$.

A positive function can have singularities on the imaginary axis (including 0 and ∞) of other types than those discussed above. It can, however, be shown (see e.g. a similar statement in Shohat and Tamarkin [32, p. 23]), that the limits

$$\lim_{z \to \infty} f(z)/z\ , \qquad \lim_{z \to i\eta} (z - i\eta)\, f(z)$$

exist for any ϵ, $0 < \epsilon < \pi/2$ if, respectively, $|\arg z|$ and $|\arg(z - i\eta)|$ do not exceed $(1-\epsilon)\pi/2$. The limits are real and non-negative.

3.2. THE MINIMUM PRINCIPLE AND SOME CONSEQUENCES

The minimum principle for harmonic functions, see e.g. [15, p. 303] can for our purpose be stated as follows:

MINIMUM PRINCIPLE. An analytic function f belongs to P', iff

A. $f(z)$ is regular for $z \in \mathbb{C}^+$,

B''. $\underline{\lim}_{z \to i\alpha} \operatorname{Re} f(z) \geq 0 \quad (-\infty \leq \alpha \leq +\infty,\ z \in \mathbb{C}^+)$.

The limit process $z \to i\alpha$ means that one should consider all possible paths along which a moving point $z \in \mathbb{C}^+$ can approach the point $i\alpha$ (also when $i\alpha = \infty$).

Remark. If $f(z)$ is defined and continuous at $i\alpha$, then $\lim \operatorname{Re} f(z) = \operatorname{Re} f(i\alpha)$. For singularities at $i\eta$ or ∞ of the form discussed in Section 3.1, condition B'' means just the inequalities (3.2'''). In particular, a pole must be simple and have a positive residue. The consideration of $i\alpha = \infty$ is important.

The reader is advised to find out why, for example, the functions $-z$, $-z^2$, z^3, $-z^{-1}$, $-z/(z^2 + 1)$, do not satisfy condition B''.

COROLLARY 1. If $g(z) = f(z) - az$ is regular at ∞, then $f(z) \in P'$ iff $a \geq 0$ and $g(z) \in P'$. A similar conclusion holds if $f(z) - a(z - i\eta)^{-1}$ or $f(z) - a(1 - i\eta z)/(z - i\eta)$ is regular at $i\eta$, $(\eta \in \mathbb{R})$. □

PROOF. First, suppose that $a \geq 0$ and $g(z) \in P'$. Then $az \in P'$, hence $f(z) = az + g(z) \in P'$.

Next, suppose that $f(z) \in P'$. Then $g(z)$ is regular in $\mathbb{C}^+$ and

$$\underline{\lim}_{z \to i\alpha} g(z) = \underline{\lim}_{z \to i\alpha} f(z) ,$$

where α is real and finite. The regularity at ∞ implies that $g(\infty)$ exists, i.e. $f(z) = az + g(\infty) + O(1)$, when $z \to \infty$. Then $a \geq 0$, by the theorem of Section 3.1. Moreover, $\operatorname{Re} g(\infty) \geq 0$, for

$$\operatorname{Re} g(\infty) = \lim_{y \to \infty} \operatorname{Re} g(|y^{-1}| + iy) = \lim_{y \to \infty} \operatorname{Re} f(|y^{-1}| + iy)$$

$$- \lim_{y \to \infty} \operatorname{Re} a(|y^{-1}| + iy) \geq \underline{\lim}_{z \to \infty} \operatorname{Re} f(z) - 0 \geq 0 .$$

Hence $g(z) \in P'$ by the minimum principle.

An analogous proof holds for the case with a pole at $i\eta$. Note that $\operatorname{Re}(z - i\eta)^{-1} = 0$ and $\operatorname{Re}(1 - i\eta z)(z - i\eta)^{-1} = 0$, when z is imaginary, $z \neq i\eta$. □

We shall formulate this corollary also for functions mapping the exterior of the unit disk into $\mathbb{C}^+$, since one can then eliminate the use of the transformation (2.7), in some applications to linear multistep methods.

COROLLARY 2. If

$$\psi(\zeta) = \varphi(\zeta) - a\,\frac{\zeta + \zeta_1}{\zeta - \zeta_1}, \qquad |\zeta_1| = 1,$$

is regular at ζ_1, then

$$\operatorname{Re} \varphi(\zeta) \geq 0 \quad \text{for} \quad |\zeta| \geq 1, \qquad \text{iff} \quad a \geq 0 \quad \text{and} \quad \operatorname{Re} \psi(\zeta) > 0 \quad \text{for} \quad |\zeta| \geq 1.$$

PROOF. Put

$$\zeta = \frac{z + 1}{a - 1}, \qquad \zeta_1 = \frac{i\eta + 1}{i\eta - 1}, \qquad \varphi(\zeta) = f(z), \qquad \psi(\zeta) = g(z),$$

and note that $\operatorname{Re} z > 0$ if $|\zeta| > 1$ and that η is real, if $|\zeta_1| = 1$. Then

$$g(z) = f(z) - a(1 - i\eta z)/(z - i\eta)$$

and the result follows from Corollary 1. □

Consider a rational function $f(z) = r(z)/s(z)$, where $|\deg r - \deg s| \leq 1$, ($\deg r$ = the degree of r). By division,

$$f(z) = az + c + g(z),$$

where $g(z) = O(z^{-1})$ when $z \to \infty$.

If $f \in P'$, then (by Corollary 1), $a \geq 0$, $c + g(z) \in P'$ and hence $\operatorname{Re} c \geq 0$. We cannot conclude that $g(z) \in P'$, unless $\operatorname{Re} c = 0$ (although conversely $g(z) \in P'$, $a \geq 0$ and $\operatorname{Re} c \geq 0$ implies $f(z) \in P'$.)

Fortunately $\operatorname{Re} c = 0$, in many interesting cases. We recall from Section 2.3 that f is called para-odd, iff $f(z) + \bar{f}(-z) = 0$, i.e. iff $f(z) + \overline{f(-\bar{z})} = 0$. It is easily verified that if f is para-odd, then a is real, $\operatorname{Re} c = 0$ and g is para-odd. It follows that f para-odd $\wedge$ $f \in P'$ $\Longleftrightarrow$ g para-odd $\wedge$ $g \in P'$, and if $g \not\equiv 0$, $1/g$ is para-odd $\wedge$ $1/g \in P'$. Hence $f_0(z) = r(z)/s(z)$ is para-odd, and belongs to P' iff the functions $f_1(z), f_2(z), \ldots$ defined by Euclid's algorithm,

$$\begin{aligned} g_n(z) &= f_n(z) - (a_n z + ib_n) \\ f_{n+1}(z) &= 1/g_n(z), \qquad \text{iff} \quad g_n \neq 0, \end{aligned} \tag{3.3}$$

will be para-odd, $a_n \geq 0$, $b_n \in \mathbb{R}$, and belong to P'. In other words, the following result has been obtained.

PROPOSITION.

$$\frac{r(z)}{s(z)} = a_0 z + ib_0 + \frac{1}{|a_1 z + ib_1} + \frac{1}{|a_2 z + ib_2} + \cdots + \frac{1}{|a_N z + ib_N}$$

with $a_n \geq 0$, $b_n \in \mathbb{R}$, iff r/s is para-odd and $r/s \in P'$. □

This result is, if we substitute $-iz$ for z, closely related to a classical result in the theory of moments, see Shohat-Tamarkin, p. 31, which also covers the case of infinite continued fractions for non-rational positive functions.

In combination with the Lemma of Section 2.3, this proposition yields necessary and sufficient conditions for a polynomial to be a Hurwitz polynomial which are equivalent to those obtained by Routh's algorithm for the case of real polynomials, Wall [34]. For complex polynomials this algorithm is due to Frank [17].

3.3. POSITIVITY CRITERIA FOR RATIONAL FUNCTIONS

Assume that the polynomials $r(z)$, $s(z)$ have no common divisor, and that one of them is not a constant. The graph $\mathcal{L} = \{q : q = r(iy)/s(iy),$ $-\infty \leq y \leq +\infty\}$ divides the q-plane or, more adequately, the Riemann sphere into sets such that the number of roots $z \in \mathbb{C}^+$ of the equation,

$$r(z) - qs(z) = 0 , \tag{3.5}$$

is constant in each of these sets. The plotting of $\mathcal{L}$ usually yields important insight into the problem characterized by r/s. In the case of linear k-step methods the interior of the region of absolute stability consists of m of these sets, $m \geq 0$. The boundary is a subset of $\mathcal{L}$, sometimes the whole of $\mathcal{L}$.

Assume that condition B" of the minimum principle (see the previous section) is satisfied. Then $\mathcal{L}$ contains no point in $q \in \mathbb{C}^-$, i.e. Eqn (3.5) has the same number of roots $z \in \mathbb{C}^-$ for all $q \in \mathbb{C}^-$. Take for example $q = -1$. Hence, if $r(z) + s(z)$ is a Hurwitz polynomial, then $r(z) - qs(z)$ is a Hurwitz polynomial for all $q \in \mathbb{C}^-$. In other words, $\operatorname{Re} r(z)/s(z) \geq 0$ for $z \in \mathbb{C}^+$, i.e. $r/s \in P'$, or since r/s is not a constant, $r/s \in P$. We therefore have the following result, well-known in circuit theory. (In [33], this characterization is even used as the definition of a positive function.)

LEMMA. A rational function $f(z) = r(z)/s(z)$ where r, s are relatively prime and at least one of them is not a constant, belongs to P, iff

A'. $r + s$ is a Hurwitz polynomial.

B"'. $\operatorname{Re} f(iy) \geq 0$ for all y, such that $s(iy) \neq 0$. □

Often the function f contains parameters, which are to be chosen to satisfy some optimality condition and perhaps also constraints other than positiveness. Then the plotting of $\mathcal{L}$ may be cumbersome or expensive, and algebraic positivity criteria may be more useful.

The simplest necessary condition is that all coefficients of r and s must have the same sign (because they belong to the closure of the set of Hurwitz polynomials). This condition is (essentially) sufficient, only if r and s are linear, see Example 1 below.

For the particular case of a para-odd function, necessary and sufficient conditions were derived in the previous section. For more general rational functions the existence of a continued fraction (3.4) with $a_n \geq 0$, $\operatorname{Im} b_n \leq 0$, $n = 0,1,\ldots,N$, is a sufficient, but not a necessary condition for positivity. Nevertheless, the ideas of the previous section, are indirectly useful in a more general situation.

First, Euclid's algorithm and Corollary 1 (of the previous section) can often be used for one or two steps to simplify $f(z)$, even if f is not para-odd, see e.g. the treatment of φ_3 in Section 4.1.

Second, condition A' in the lemma of this section, can be investigated by Routh's algorithm.

Third, condition B''' can be written,

$$p(y) \geq 0 \quad \text{for} \quad y \in \mathbb{R} \qquad \text{where} \qquad p(y) = \operatorname{Re} r(iy)\,\overline{s(iy)}\,, \quad (3.6)$$

$p(y)$ is a real polynomial. A systematic way of investigating this is by means of Sturm chains, see e.g. [18, p.175], which is again an application of Euclid's algorithm, this time to the function $p(y)/p'(y)$. If r/s is a real function, p becomes an even function, and this fact simplifies the investigation. The calculations can be organized in a modified Routh tableau, see Šiljak [33].

There is another approach to positivity criteria, see Jeltsch [26]. Let $f(z) = r(z)/s(z)$. Suppose that for every imaginary q there are no roots of Eqn (3.5) in $\mathbb{C}^+$. (This hypothesis can be investigated by the complex variant of Routh's algorithm.) This means that $f(\mathbb{C}^+)$ contains no point on the imaginary axis, hence either $f(\mathbb{C}^+) \subset \mathbb{C}^+$ or $f(\mathbb{C}^+) \subset \mathbb{C}^-$. In order to exclude the latter case, it is sufficient to verify that $f(z_0) \in \mathbb{C}^+$ for at least one $z_0 \in \mathbb{C}^+$.

Similar criteria can be developed for rational functions satisfying the condition $\operatorname{Re} \varphi(\zeta) \geq 0$ on the unit disk (or its exterior). The Routh algorithm is then to be replaced by the Schur-Cohn algorithm. Nevertheless, it may be advantageous to transform the circle case into the half-plane case for other reasons, e.g. the simple necessary condition mentioned earlier and the convenient handling of certain symmetry conditions. The transformation (2.7) is much simpler to perform than is generally believed. In fact, the coefficients in the polynomial

$$r(z) \equiv r_k(z) \equiv (z-1)^k \rho((z+1)/(z-1)),$$

where $\rho(\zeta) = \sum_{j=0}^{k} \alpha_j \zeta^j$, can be calculated by the following algorithm, similar to synthetic division,

$$p_0(z) \equiv 1, \qquad r_0(z) \equiv \alpha_k \tag{3.7'}$$

$$p_i(z) \equiv (z-1)\cdot p_{i-1}(z), \qquad (i = 1,2,\ldots,k), \tag{3.7''}$$

$$r_i(z) \equiv (z+1)\cdot r_{i-1}(z) + \alpha_{k-i}\, p_i(z). \tag{3.7'''}$$

We shall now investigate a few simple examples.

<u>Example 1</u>. Let a, b, c, d be real, $|c| + |d| \neq 0$, $ad - bc \neq 0$ and let

$$f(z) = \frac{az+b}{cz+d}.$$

When is $f \in P'$? We investigate condition (3.6), which is equivalent to B''',

$$p(y) = \operatorname{Re}(aiy+b)(-ciy+d) = bd + acy^2.$$

$$p(y) \geq 0 \iff bd \geq 0,\ ac \geq 0.$$

Condition A' $\iff (a+c)z + (b+d) \in$ Hurwitz $\iff$

$$(a+c)(b+d) > 0.$$

These conditions are satisfied iff a, b, c, d have the same sign, and $a + c \neq 0$ and $b + d \neq 0$. The latter conditions follow from the assumption that $ad - bc \neq 0$, and can be ignored if it is acceptable that $f(z)$ degenerates into a constant.

<u>Example 2</u>. Let a_0, b_0, b_1, b_2 be real, $b_2 \neq 0$,

$$f(z) = \frac{b_2z^2 + b_1z + b_0}{z + a_0}.$$

When is $f \in P$?

$$f(z) = b_2z + \frac{(b_1 - b_2a_0)z + b_0}{z + a_0}$$

By Corollary 1 and Example 1, we obtain the necessary and sufficient conditions for $f \in P$ (since $f \not\equiv 0$):

$$b_2 > 0, \qquad b_1 - b_2a_0 \geq 0, \qquad b_0 \geq 0, \qquad a_0 \geq 0.$$

Example 3. Let a_0, a_1, b_0, b_1 be real, and let

$$f(z) = \frac{z^2 + b_1 z + b_0}{z^2 + a_1 z + a_0}$$

When is $f \in P$?

We have the necessary conditions

$$a_0 \geq 0, \qquad a_1 \geq 0, \qquad b_0 \geq , \qquad b_1 \geq 0 . \tag{3.8'}$$

Condition A' then gives the inequalities

$$b_0 + a_0 > 0, \qquad b_1 + a_1 > 0 . \tag{3.8''}$$

We investigate (3.6)

$$\begin{aligned} p(y) &= \operatorname{Re}(-y^2 + b_1 iy + b_0)(-y^2 - a_1 iy + a_0) \\ &= y^4 + (a_1 b_1 - a_0 - b_0)y^2 + a_0 b_0 \end{aligned}$$

Since we already have the condition $a_0 b_0 \geq 0$, we find that $p(y) \geq 0$ yields the new condition

$$a_1 b_1 - a_0 - b_0 \geq 0 \vee a_0 b_0 - (a_1 b_1 - a_0 - b_0)^2/4 \geq 0 , \tag{3.8'''}$$

in addition to (3.8') and (3.8").

4. THE USE OF POSITIVE FUNCTIONS IN INVESTIGATION OF NUMERICAL STABILITY

4.1. THE FORMULATION OF BOUNDARY CONDITIONS FOR A METHOD OF LINES-SOLUTION TO A HYPERBOLIC PROBLEM

The following example comes from a joint work with G. Sköllermo in 1977 (not yet published). Consider, as a model problem for a hyperbolic system, the equation

$$\frac{\partial u}{\partial t} = \frac{\partial u}{\partial x} . \qquad u(x,0) \text{ and } u(1,t) \text{ are given.} \tag{4.1}$$

Let $x_n = nh$, $n = 0,1,2,\ldots,N$, $Nh = 1$. $u_n(t)$ is intended to be an approximation of $u(x_n,t)$. The time derivative is denoted by $\dot{u}_n$.

This PDE will be replaced by a system of ODE's, by the application of Simpson's rule to the left-hand side in the equation

$$\int_{x_{n-1}}^{x_{n+1}} \frac{\partial u}{\partial t}\, dx = \int_{x_{n-1}}^{x_{n+1}} \frac{\partial u}{\partial x}\, dx, \qquad (n = 1,2,\ldots,N-1)$$

Hence,

$$\frac{h}{3}(\dot{u}_{n-1} + 4\dot{u}_n + \dot{u}_{n+1}) = u_{n+1} - u_{n-1} \tag{4.2}$$

$u_N(t)$ is given by the boundary condition, but we need also an equation for $u_0(t)$, which is not given for the PDE; $x = 0$ is a point of outflow in this model problem. Since Simpson's formula is fourth order accurate, we would wish at least a third order accurate equation for u_0 such that the differential system is <u>stable</u> and still has a simple structure. We choose the equation

$$\frac{h}{3}\left(\frac{5}{2}\dot{u}_0 + 4\dot{u}_1 - \frac{1}{2}\dot{u}_2\right) = 2(u_1 - u_0) \tag{4.3}$$

which is

$$2\int_{x_0}^{x_1} \frac{\partial u}{\partial t}\,dx + O(h^4) = 2\int_{x_0}^{x_1} \frac{\partial u}{\partial x}\,dx .$$

Let $u = (u_0, u_1, \dots, u_{N-1})^T$. We then obtain an almost tridiagonal $N \times N$ system of the form,

$$\frac{h}{3} B\dot{u} = Au . \tag{4.4}$$

(The first row has three non-zero elements instead of two.)

This has solutions of the form $e^{\lambda t}$ if

$$D_N(z) \equiv \det(Bz - A) = 0, \qquad z = h\lambda/3 \tag{4.5}$$

$$D_N(z) = \begin{vmatrix} \frac{5}{2}z + 2 & 4z - 2 & -\frac{1}{2}z & 0 & \cdots & 0 & 0 \\ z + 1 & 4z & z - 1 & 0 & \cdots & 0 & 0 \\ 0 & z + 1 & 4z & z-1 & \cdots & 0 & 0 \\ \cdots & \cdots & \cdots & \cdots & \cdots & \cdots & \cdots \\ 0 & 0 & 0 & 0 & \cdots & 4z & z-1 \\ 0 & 0 & 0 & 0 & \cdots & z+1 & 4z \end{vmatrix} \tag{4.6}$$

It is easily verified that

$$D_{n+1}(z) = 4z\,D_n(z) - (z^2 - 1)\,D_{n-1}(z) , \qquad n \geq 3 . \tag{4.7}$$

If it turns out that $D_n(z) \neq 0$ and that imaginary zeros are simple then for each n, all solutions of the ODE's are bounded, and any A-stable method can be used for the integration in time. Put

$$\varphi_n = D_n/D_{n-1} \tag{4.8}$$

Then, by (4.7) and (4.6)

$$\varphi_{n+1}(z) = 4z - (z^2-1)/\varphi_n(z), \qquad n \geq 3 \tag{4.9}$$

$$\varphi_3(z) = \frac{D_3(z)}{D_2(z)} = \frac{21\,z^3 + 21z^2 + 10z + 2}{6z^2 + 6z + 2} \tag{4.10}$$

Since $D_1(z)$ has its only zero in the left half-plane, it is sufficient (though not necessary) to show that $\varphi_n(z) \in P$ for $n \geq 3$ to show that $D_n(z) \neq 0$ for $z \in \mathbb{C}^+$ and that there are no multiple zeros on the imaginary axis.

$$\varphi_3(z) = \frac{21z}{6} + \frac{3z + 2}{6z^2 + 6z + 2}$$

Since

$$f(z) := \frac{6z^2 + 6z + 2}{3z + 2} = 2z + \frac{2z + 2}{3z + 2} \in P,$$

by Example 1 in Section 3.3, it follows that $f(z) \in P$ and hence $\varphi_3(z) \in P$.

We shall apply the minimum principle in a proof by induction applied to (4.9). Suppose that $\varphi_n(z) \in P$.

$$\varphi_n \in P \implies 1/\varphi_n \in P \implies \varphi_{n+1} \text{ regular in } \mathbb{C}^+.$$

$$\varphi_n \in P \implies 1/\varphi_n \in P \implies \operatorname{Re} \varphi_{n+1}(iy) \geq 0$$

at regular imaginary points. If $\varphi_n(i\eta) = 0$, η real and finite, then, since $1/\varphi_n \in P$, the residue of $1/\varphi_n$ at $i\eta$ is positive, say $r > 0$. The residue of φ_{n+1} at $i\eta$ is then $(1 + \eta^2)r > 0$.

It only remains to look at the behavior at ∞. Let

$$a_j = \lim_{z \to \infty} \varphi_j(z)/z, \qquad j = 3,4,5,\ldots$$

By (4.9) and (4.10),

$$a_{n+1} = 4 - 1/a_n,$$

$$a_3 = 21/6.$$

The roots of the equation $X = 4 - 1/X$ are $2 \pm 3^{1/2}$. Since a_3 lies between the two roots, a_n will increase towards the larger root. Hence $a_N > 0$, for all N.

It then follows that $\varphi_{n+1} \in P$, $n \geq 2$, and hence $D_n(z) \neq 0$ in $\mathbb{C}^+$ for $n \geq 2$, which was to be proved.

4.2. A-STABILITY AND ACCURACY FOR LINEAR MULTISTEP METHODS

In section 2.1, linear multistep methods were considered for differential equations of the form,

$$d^q y/dt^q = f(t,y), \qquad q = 1,2.$$

The notion of the canonical function was introduced. Let

$$g(z) = (s(z))/r(z))^{1/q}.$$

It was shown that A-stability (for $q = 1$) or unconditional stability (for $q = 2$) was characterized by the statement,

$$g(z) \in P . \tag{4.11}$$

By Taylor's formula, one can show [25] that there exist constraints c, p, independent of the function y, such that

$$\sum_{j=0}^{k} \alpha_j y(t_{n+j}) - h^q \sum \beta_j y^{(q)}(t_{n+j}) \sim ch^{p+q} y^{(p+q)}(t_n) \quad .$$

If we choose $y(t) = e^t$, and put $\zeta = e^h$, and perform the transformation (2.7), we obtain, after a straight-forward calculation, see e.g. [8], [12] [25],

$$g(z) = \log\left(\frac{z+1}{z-1}\right)^{-1} + c'\left(\frac{z}{2}\right)^{1-p} + \mathcal{O}(z^{-p}) \quad . \tag{4.12}$$

where $c' = c/(q\sigma(1))$. If we expand the logarithm, we obtain,

$$g(z) = z/2 - z^{-1}/6 + c'(z/2)^{1-p} + \mathcal{O}(z^{-2}) \tag{4.13}$$

If $p > 2$, this implies,

$$g(z) - z/2 \sim -z^{-1}/6 . \qquad (z \to \infty) \tag{4.14}$$

If (4.11) is true, then the left-hand side belongs to P', by Corollary 1 of Section 3.2, but by (3.2'''), a function in P' can behave like az^{-1} only if $a \geq 0$. The contradiction with (4.14) shows that $p \leq 2$. The same argument applied to (4.13) shows that if $p = 2$ then $c' \geq 1/12$.

Moreover, if $p = 2$, $c' = 1/12$, then by (4.13) either $g(z) - z/2 \equiv 0$ or

$$g(z) - z/2 \sim az^{-\alpha}, \qquad \alpha \geq 2, \quad a \neq 0 .$$

By (3.2'''), the latter alternative is incompatible with positivity. Hence if $p = 2$, $c' = 1/12$, we have $g(z) = z/2$, i.e.

$$s(z)/r(z) = (z/2)^q \quad . \tag{4.15}$$

If $q = 1$ this defines the trapezoidal method [9]. If $q = 2$ we obtain a method that is equivalent to the application of the trapezoidal rule [12] to the first order system,

$$y' = z , \qquad z' = f(t,y).$$

4.3. STABILITY OF LINEAR MULTISTEP METHODS ON A CLASS OF STIFF NON-LINEAR PROBLEMS

Consider an initial value problem for a first order system of s complex differential equations

$$dy/dt = f(t,y) . \tag{4.16}$$

Let $y'(t)$ and $y''(t)$ be two arbitrary solutions of (4.16) with different initial conditions. Let $\langle\cdot,\cdot\rangle$ be an inner product in $\mathbb{C}^s$, and $\|\cdot\|$ be the associated norm. Assume that the system (4.16) is <u>contractive</u> in the sense that

$$d\|y'(t) - y''(t)\|^2/dt \le 0 \qquad \forall t .$$

This condition is natural for stiff problems. It is equivalent to the condition

$$\text{Re}\,\langle f(y') - f(y''), y' - y''\rangle \le 0 , \qquad \forall y', y'' \text{ in } \mathbb{C}^s. \tag{4.17}$$

In [11] we showed that the linear k-step method (or rather the corresponding "one-leg method") with the canonical function $r(z)/s(z)$ has a similar contractivity property (G-stability), if there exist polynomials

$$\{\varphi_i(z)\}_{i=0}^{k-1}, \qquad \{\psi_i(z)\}_{i=0}^{k} ,$$

such that

$$\text{Re}\,\overline{s(z)}\, r(z) = \text{Re}\, z \cdot \sum_{i=0}^{k-1} |\varphi_i(z)|^2 + \sum_{i=0}^{k} |\psi_i(z)|^2 . \tag{4.18}$$

The coefficients φ_{ij} of the φ_i are used for defining a norm in $\mathbb{C}^{ks}$. Let $V = (v_1, v_2, \dots , v_k)$ when $v_j \in \mathbb{C}^s$, then we define

$$\|V\|^2 = \sum_{i=0}^{k-1} \|\sum_{j=0}^{k-1} \varphi_{ij} v_{j+1}\|^2 .$$

This is a norm, if the polynomials $\varphi_i(z)$ are linearly independent. Let y'_n, y''_n be vector sequences obtained when applying a one-leg method to a contractive system (4.16), and put

$$Y'_n = (y'_n, y'_{n+1}, \dots , y'_{n+k-1}) .$$

It is shown in [11] that if there exists an expansion of the form (4.18) for the method, then

$$\|Y'_{n+1} - Y''_{n+1}\| \leq \|Y'_n - Y''_n\|, \qquad n = 0,1,2,\ldots \tag{4.19}$$

This is the contractivity property (G-stability) mentioned above.

THEOREM. If $s(z)/r(z)$ is a positive function, there exists an expansion of the form (4.18), where the polynomials φ_i, $i = 0,1,2,\ldots,k-1$, are linearly independent, i.e. A-stability implies G-stability.

We shall sketch the proof of this result which is thought to be new. A detailed treatment of this is given in [13].

Let $r_0(z) = r(z)$, $s_0(z) = s(z)$ and suppose that $s_0/r_0 \in P$. We shall construct a sequence of positive functions, $s_n(z)/r_n(z)$, $n = 1,2,\ldots$ of decreasing degree. Suppose that $s_{n-1}(z)/r_{n-1}(z) \in P$.

If $r_{n-1}(z)$ has an imaginary zero, say, at $i\eta_n$, $-\infty \leq \eta_n \leq +\infty$, then let

$$r_n(z) = \frac{r_{n-1}(z)}{z - i\eta_n} ,$$

and let

$$\frac{s_n(z)}{r_n(z)} = \frac{s_{n-1}(z)}{r_{n-1}(z)} - a_n \frac{1 - i\eta_n z}{z - i\eta_n} , \tag{4.20}$$

where a_n is chosen such that $s_n(z)/r_n(z)$ becomes regular at $i\eta_n$. Then by Corollary 1 of Section 3.2, $a_n > 0$ and $s_n(z)/r_n(z) \in P'$, i.e. the function either belongs to P or it is a constant. In the latter case the construction is finished.

Otherwise, i.e. if $r_{n-1}(iy) \neq 0$, $\forall y \in \mathbb{R}$, let $i\eta_n$ be one of the points where $\operatorname{Re} s_{n-1}(z)/r_{n-1}(z)$ assumes it infimum over $\mathbb{C}^+$. Put

$$\alpha_n + i\beta_n = s_{n-1}(i\eta_n)/r_{n-1}(i\eta_n) .$$

Then let

$$\frac{s_n(z)}{r_n(z)} = \frac{r_{n-1}(z)}{s_{n-1}(z) - (\alpha_n + i\beta_n)\, r_{n-1}(z)} - a_n \frac{1 - i\eta_n z}{z - i\eta_n} \tag{4.21}$$

It is left as an exercise to the reader to verify that the first term on the right-hand side is a positive function with a pole at $i\eta_n$. Again, $r_n(z)$ is the ratio of two denominators on the right-hand side, and a_n is chosen so that s_n/r_n becomes regular at $i\eta_n$. By Corollary 1 of Section 3.2, $a_n > 0$ and $s_n(z)/r_n(z) \in P'$ etc. as in the previous case.

By a straightforward computation, (4.21) leads to

$$|z - i\eta_n|^2 \operatorname{Re} s_n(z)\, \overline{r_n(z)}$$
$$= \operatorname{Re} \overline{s_{n-1}(z)}\, r_{n-1}(z) - \alpha_n |r_{n-1}(z)|^2 - a_n(1+\eta_n^2) \operatorname{Re} z \cdot |r_n(z)|^2$$

Similarly, in the first alternative, (4.20) yields the same formula with $\alpha_n = 0$. In both cases this formula gives a downwards recurrence formula for proving an expansion of the form (4.18). The reader is referred to [13] for a proof that the φ_i become linearly independent.

Since this expansion is a consequence of the positivity assumption alone (e.g. consistency is not assumed) there are generalizations of this result to the other stability requirements mentioned in Section 2.1, see [13].

The construction of the sequence $\{s_n/r_n\}$ is similar to one employed in the Brune synthesis of passive circuits [4 , p. 143]. It is not known to the author if the expansion (4.18) has any circuit theoretical interpretation.

5. INTEGRAL REPRESENTATION OF POSITIVE FUNCTIONS

THEOREM 1. $f(z) \in P'$, iff there exist constants $a \in \mathbb{R}^+$, $b \in \mathbb{R}$ and a non-decreasing and bounded function σ, such that

$$f(z) = az + ib + \int_{-\infty}^{\infty} \frac{1 - itz}{z - it}\, d\sigma(t) \,. \quad \square \tag{5.1}$$

For a proof of this theorem, see [1, p. 7] or [32, p. 23]. In the latter source it is also proved that $f(z)/z \to a$, as $z \to \infty$ in any sector,

$$|\arg z| < \pi/2 - \epsilon,\ 0 < \epsilon < \pi/2 \,.$$

In [1] the following variant is also proved.

THEOREM 2. $f(z) \in P'$ and $f(x) = O(x^{-1})$ when $x \to +\infty$, iff there exists a non-decreasing and bounded function α such that

$$f(z) = \int_{-\infty}^{\infty} (z - it)^{-1}\, d\alpha(t) \,. \quad \square \tag{5.2}$$

The theorems are based on a similar representation given in 1911 by Herglotz and F. Riesz for functions with non-negative real parts in the unit circle. The half-plane case is due to R. Nevanlinna. The representations are based on the Poisson formula for the solution of the Dirichlet problem on a half-plane. In fact,

$$\frac{1}{2}(\alpha(t-0)+\alpha(t+0)) = \text{const} + \lim_{x\to 0+} \frac{1}{\pi}\int_0^t \operatorname{Re} f(x+iy)dy .$$

If f is regular at the point it, then

$$\alpha'(t) = \frac{1}{\pi} \operatorname{Re} f(it)$$

If f has a pole at it, then $\alpha(t)$ has a jump (point mass) equal to the residue. The relation between α and σ is given by

$$d\alpha(t) = (1+t^2)\, d\sigma(t) .$$

Aronszajn and Donoghue [2] list a large number of facts; related to (5.1) and (5.2). The representation (5.1) describes the class P' as a convex cone generated by the constants $\pm i$ and the functions which are used to describe the principal part at poles. The representation formulas are simple to apply. In fact the first proof [9] of the upper bound for the order of an A-stable linear multistep method with canonical function $r(z)/s(z)$ was based on (5.2) which for $f(x) = r(x)/s(x)$, gives, after multiplication by x,

$$x\frac{r(x)}{s(x)} = \int_{-\infty}^{\infty} \frac{x^2}{x^2+t^2}\, d\alpha(t), \qquad x > 0 .$$

This shows that $xr(x)/s(x)$ is a non-decreasing function, since $x^2/(x^2+t^2)$ is so for each t. On the other side, let the order of accuracy of the method be p. Then (compare (4.12)),

$$\frac{r(z)}{s(z)} - \log\frac{z+1}{z-1} = O(z^{-p-1}) .$$

Hence if $p \geq 3$

$$\frac{xr(x)}{s(x)} = 2 + \frac{2}{3}x^{-2} + O(x^{-3}) ,$$

where the right-hand side is clearly decreasing for large x. The contradiction shows that $p \leq 2$.

6. EXTENSIONS OF THE CONCEPT OF POSITIVE FUNCTIONS

6.1. POSITIVE MATRIX FUNCTIONS

Let $A(z)$ be an $n\times n$ matrix, in general not Hermitean, the elements of which are analytic in $\mathbb{C}^+$. We say that

$$A(z) \in P_n \quad \text{iff} \quad u^H A(z)u \in P, \quad \forall u \in \mathbb{C}^n, \quad u \neq 0 , \tag{6.1}$$

$$A(z) \in P'_n \quad \text{iff} \quad u^H A(z)u \in P', \ \forall u \in \mathbb{C}^n . \tag{6.1'}$$

The class P'_n plays an important role in the design of passive networks, see Belevitch [4, Ch. 7]. $A(z)$ is called a positive matrix function if $A(z) \in P'_n$. A recent application of a related concept is found in Genin et al. [22]. The writer does not yet know of any successful applications to numerical analysis. It is hoped that this presentation can stimulate future applications.

Some properties are directly generalized from the scalar case, e.g.

$$A_i(z) \in P'_n,\ \alpha_i \geq 0 \implies \sum \alpha_i A_i(z) \in P'_n \tag{6.2}$$

$$A(z) \in P'_n,\ f(z) \in P' \implies A(f(z)) \in P'_n \ . \tag{6.3}$$

A corresponding result for the composition in reverse order reads,

$$A(z) \in P_n,\ f(z) \in P'_n \implies f(A(z)) \in P'_n \ , \tag{6.4}$$

(An improvement is given below.)

In order to show this, we first consider some particular cases. We write, for the sake of brevity, A instead of $A(z)$.

LEMMA. Let $A \in P_n$ and let t be a real constant. Then the matrix functions A^{-1}, $A - itI$, $(I - itA)(A - itI)^{-1}$ also belong to P_n.

PROOF. A^{-1} exists, because otherwise there would exist a vector u, $u \neq 0$, such that $Au = 0$, hence $u^H Au = 0$, which contradicts the definition. Consider $u^H A^{-1} u$ and put $u = Av$. Then

$$\text{Re}\ u^H A^{-1} u = \text{Re}\ v^H Av > 0 \ ,$$

hence $A^{-1} \in P_n$. Moreover, if $t \in \mathbb{R}$,

$$\text{Re}\ u^H (A - itIu) = \text{Re}\ u^H Au - \text{Re}\ it\ u^H u > 0 \ ,$$

hence $(A - itI) \in P_n$. By combination of these results $(A - itI)^{-1} \in P_n$. The final results then follow from the identity

$$(I - itA)(A - itI)^{-1} = (1 + t^2)(A - itI)^{-1} - itI \in P_n . \qquad \square$$

In order to derive (6.4), we shall use the Herglotz type representation (5.1).

$$f(z) = az + ib + \int_{-\infty}^{\infty} \frac{1 - itz}{z - it}\, d\sigma(t), \qquad z \in \mathbb{C}^+ \ . \tag{6.5}$$

It follows directly from (6.1) that the spectrum of $A(z)$ lies in $\mathbb{C}^+$. Therefore, we can write

$$f(A) = aA + ibI + \int_{-\infty}^{\infty} (I - itA)(A - it)^{-1}\, d\sigma(t) \ , \tag{6.6}$$

and since by the Lemma the right-hand side is a positive combination of positive matrix functions (a limiting case of (6.2)), $f(A(z)) \in P'_n$.

Some of the ideas in Chapter 3 can also be generalized. The matrix function $A(z)$ has a pole at ti, if at least one of its elements has a pole there. Then there exists a constant matrix R, such that

$$B(z) = A(z) - \frac{1 - itz}{z - it} R \tag{6.7}$$

is regular at ti. Let $A(z) \in P'_n$. Then for any $u \in \mathbb{C}^n$, the residue of $u^H A(z)u$, i.e. $(1 + t^2)\, u^H Ru$, must be real and positive, by Section 3.1 (Theorem). It follows that R <u>is Hermitean, positive-semi-definite</u>. Similarly, by Section 3.2 (Corollary 1), it follows that $u^H B(z)u \in P'$, i.e. $B(z) \in P'_n$.

Criteria for positivity of matrix functions are treated in Šiljak [33].

Equation (6.5) is not the only extension of the Herglotz type representation to the matrix case. The matrices A_∞, B_∞, $H(t)$ in the formula,

$$A(z) = A_\infty z + iB_\infty + \int_{-\infty}^{\infty} \frac{1 - itz}{z - it}\, dH(t) , \qquad z \in \mathbb{C}^+ , \tag{6.8}$$

can be obtained from the following identity (where e_j, e_k are basic unit vectors)

$$\begin{aligned} 4e_j^T A e_k = {} & (e_j + e_k)^H A(e_j + e_k) - (e_j - e_k)^H A(e_j + e_k) \\ & - i(e_j + ie_k)^H A(e_j + ie_k) + i(e_j - ie_k)^H A(e_j - ie_k), \end{aligned}$$

and applying the scalar Herglotz representation (6.5) to each of the terms on the right-hand side.

Now consider any $u \in \mathbb{C}^n$ and put $f(z) = u^H A(z)u$ into (6.5), and identify with (6.8).

$$u^H A_\infty u = a(u) \geq 0, \qquad u^H B_\infty u = b(u) \in \mathbb{R}, \qquad u^H dH(t) u = d\sigma(t,u) \geq 0 .$$

It follows that in (6.8), A_∞ <u>and</u> $dH(t)$ <u>are positive semi-definite Hermitean matrices, while</u> B_∞ <u>is just Hermitean.</u>

We shall use (6.8) for deriving a remarkable result, which for the case of rational functions, is proved by an entirely different method in [4, Ch. 7, Theorem 10].

<u>THEOREM</u>. Let $A(z) \in P'_n$. Then the null-space (and hence the rank) of $A(z)$ is the same for all $z \in \mathbb{C}^+$. The range of $A(z)$ is the unitary complement of the null-space.

PROOF. To begin with assume only that there exist $z_0 \in \mathbb{C}^+$ and $u \in \mathbb{C}^n$ such that

$$\operatorname{Re} u^H A(z_0) u = 0 . \tag{6.9}$$

Then by (6.8) since A_∞, B_∞, $dH(t)$ are Hermitean,

$$0 = u^H A_\infty u \cdot \operatorname{Re} z_0 + 0 + \int_{-\infty}^{\infty} \operatorname{Re} \frac{1 - itz_0}{z_0 - it} \, d(u^H H(t) u) .$$

Since $\operatorname{Re} z_0 > 0$, $\operatorname{Re}(1 - itz_0)/(z_0 - it) > 0$, we conclude that

$$u^H A_\infty u = 0, \qquad u^H dH(t) u = 0, \qquad \forall t .$$

Since A_∞ and $dH(t)$ are positive semi-definite, it follows that

$$A_\infty u = 0, \qquad dH(t) u = 0, \qquad \forall t.$$

Then, by (6.8) for all $z \in \mathbb{C}^+$

$$A(z)u = iB_\infty u, \qquad A(z)^H u = -iB_\infty u, \tag{6.10}$$

where B_∞ as before is a constant Hermitean matrix.

We now strengthen assumption (6.9), and assume that

$$A(z_0)u = 0. \tag{6.11}$$

It then follows from (6.10) that $B_\infty u = 0$ and hence

$$A(z)u = (A(z))^H u = 0, \qquad \forall z \in \mathbb{C}^+ . \tag{6.12}$$

In other words, if u belongs to the null-space of $A(z_0)$, for a particular $z_0 \in \mathbb{C}^+$, it belongs to the null-space of $A(z)$ for all $z \in \mathbb{C}^+$.

It also follows that the null-space of $(A(z))^H$ contains the null-space of $A(z)$. If we had worked with $(A(z))^H$ instead of $A(z)$ we would have obtained the opposite inclusion. Hence $A(z)$ and $A(z)^H$ have the same null-spaces. The range statment now follows because the range of A is always the unitary complement of the null-space of A^H. □

<u>COROLLARY 1</u>. Let $A(z) \in P'_n$. If for some $z_0 \in \mathbb{C}^+$, $A(z_0)$ has an imaginary eigenvalue ti, then ti is an eigenvalue of $A(z)$ for all $z \in \mathbb{C}^+$, and the corresponding eigenvectors are independent of z. (Substitute $A(z) - itI$ for $A(z)$) in the theorem.)

COROLLARY 2. Let $A(z) \in P'_n$. There exists a constant unitary matrix U that transforms $A(z)$ to the form

$$\begin{bmatrix} iT & 0 \\ 0 & A_1(z) \end{bmatrix}$$

where $iT = \operatorname{diag}(it_j)$ contains the (constant) imaginary eigenvalues of $A(z)$. $A_1(z) \in P'$ has its spectrum in $\mathbb{C}^+$.

We can now generalize (6.4). Assume that $A(z) \in P'_n$ and that the limits $f(it_j) = \lim_{x \downarrow 0} f(x + it_j)$ exist and are finite. Then

$$f(A(z)) = U \begin{bmatrix} f(iT) & 0 \\ 0 & f(A_1(z)) \end{bmatrix} U^H$$

The same kind of proof as was used for (6.4) shows that $f(A_1(z)) \in P'_n$, while $f(iT) \in P'_n$, because $\operatorname{Re} f(it_j) \geq 0$. Hence $f(A(z)) \in P'_n$, and the assumption concerning $A(z)$ in (6.4) can be relaxed to the requirement that $A(z) \in P'_n$ and that the imaginary eigenvalues are only points ti for which $\lim_{x \downarrow 0} f(x + ti)$ exists and is finite. (A simple example of the exceptional case is when $A(z)$ is singular and $f(z) = z^{-1}$.)

6.2. POSITIVE ALGEBRAIC FUNCTIONS

Equations of the form,

$$\sum_{i=0}^{n} \sum_{j=0}^{k} a_{ij} p^{k-j} q^i = 0 ,$$

are encountered in the study of linear multistep multiderivative methods [21] and other methods for the numerical solution of first order systems of ODEs. This equation defines an n-valued algebraic function $q(p)$. The A-stability of the method is expressed by the condition that $\operatorname{Re} p > 0 \implies \operatorname{Re} q < 0$, for each branch of $q(p)$. Then $-q(p)$ is called a positive algebraic function. This is an extension of the class P defined in Chapter 1. Positive algebraic functions have no zeros or poles in $\mathbb{C}^+$, but branch points are allowed in $\mathbb{C}^+$.

An alternative condition for A-stability is that $-p(q)$ is a k-valued positive algebraic function.

The reader is referred to Genin [21] for more information of these functions and applications. One has to note that Genin's definition of the 'error order' is not equivalent with the one commonly used, so that some of his conclusions have to be modified, see Jeltsch [38] and Wanner et al. [35].

6.3. FUNCTIONS OF SEVERAL VARIABLES

In a study of an implicit alternating direction scheme for hyperbolic and parabolic equations, Beam and Warming [37] encountered a question concerning the positiveness of a function of two independent variables,

$$f(z_1,z_2) = \frac{z_1 + z_2}{1 + az_1z_2}, \qquad z_j \in \mathbb{C}^+, \quad a > 0 .$$

We rewrite this equation in the form,

$$\frac{1}{f(z_1,z_2)} = \frac{1}{z_1 + z_2} + \frac{a}{1/z_2 + 1/z_1} .$$

Then it follows that $\operatorname{Re} f(z_1,z_2) > 0$ for $z_j \in \mathbb{C}^+$, from the basic facts (see Chapter 1) that if $w_1 \in \mathbb{C}^+$, $w_2 \in \mathbb{C}^+$, then $w_1 + w_2 \in \mathbb{C}^+$ and $1/w_1 \in \mathbb{C}^+$. This extension does not seem to have been considered much yet.

6.4. POSITIVE FUNCTIONS OUTSIDE THE UNIT DISK

The linear multistep method (2.1) can be written in operator form, where $Ey_n = y_{n+1}$,

$$\rho(E)y_n = h\sigma(E)\, f(y_n,t_n) , \tag{6.13}$$

and the A-stability condition (2.4) reads,

$$\operatorname{Re} \rho(\zeta)/\sigma(\zeta) > 0 \qquad \text{for} \quad |\zeta| > 1 . \tag{6.14}$$

Usually it is advantageous to perform the transformation (2.7) to the half-plane case. In the study of ℓ_2-estimates of the error in applications to stiff differential equations, Odeh and Liniger [30], Dahlquist and Nevanlinna [14], it turned out however to be better to use (6.14) directly. This was due to the following interesting consequence of (6.14), see also [39, p. 18]. Let $\langle \cdot,\cdot \rangle$ be an inner-product in $\mathbb{C}^s$, and let the operator $\rho(E)/\sigma(E)$ be defined by the Laurent expansion around ∞,

$$\rho(\zeta)/\sigma(\zeta) = \sum_{n=0}^{\infty} \gamma_n \zeta^{-n} .$$

Then

$$\operatorname{Re} \sum_{n=0}^{\infty} R^{-2n} \left\langle u_n, \frac{\rho(E)}{\sigma(E)} u_n \right\rangle > 0, \qquad R > 1 , \tag{6.15}$$

for any non-trivial sequence $\{u_n\}$ of vectors in $\mathbb{C}^s$, such that $u_n = 0$ for $n < 0$ and such that the series

$$\hat{u}(\zeta) = \sum_{n=0}^{\infty} u_n \zeta^{-n} , \tag{6.16}$$

converges for $|\zeta| \geq R$.

Let $C = \{\zeta : |\zeta| = R\}$. One obtains (6.15) by substituting (6.16) into the left-hand side of the relation

$$\operatorname{Re} \frac{1}{2\pi i} \int_C \left\langle \hat{u}(\zeta), \frac{\rho(\zeta)}{\sigma(\zeta)} \hat{u}(\zeta) \right\rangle \frac{d\zeta}{\zeta} \geq \min \operatorname{Re} \frac{\rho(\zeta)}{\sigma(\zeta)} \cdot \frac{1}{2\pi i} \int_C \|u(\zeta)\|^2 \frac{d\zeta}{\zeta} > 0.$$

In [30] and [14], u_n was chosen to be equal to the difference of two perturbed solutions of (6.13). Then (6.15) yielded bounds for $\sum_{n=0}^{N} \|u_n\|^2 R^{-2n}$, $R \geq 1$, for systems satisfying contractivity assumptions similar to (4.17).

REFERENCES

1. N. I. Akhiezer and I. M. Glazman (1963), Theory of linear operators in Hilbert Space, vol. II, Fredrick Ungar Publ. Co., New York.

2. N. Aronszajn and W. F. Donoghue, Jr. (1957), On exponential representations of analytic functions in the upper half-plane with positive imaginary part, Journal d'Analyse Mathématique, vol. 5 321-388.

3. S. Barnett and D. D. Šiljak (1977), Routh algorithm: a centennial survey, SIAM Rev. 19, 472-489.

4. V. Belevitch (1968), Classical network theory, Holden-Day, San Francisco.

5. T. A. Bickart and E. I. Jury (1978), Arithmetic tests for A-stability, A(α)-stability and stiff stability, BIT, 18 9-21.

6. O. Brune (1931), Synthesis of a finite two-terminal network whose driving point impedance is a prescribed function of frequency, J. Math. Phys., 10, 191-236.

7. C. W. Cryer (1973), A new class of highly-stable methods; A_0-stable methods, BIT 13, 153-159.

8. G. G. Dahlquist (1956), Convergence and stability in the numerical integration of ordinary differential equations, Math. Scand 4, 33-53.

9. _____ (1963), A special stability problem for linear multistep methods, BIT 3, 27-43.

10. _____ (1975), Error analysis for a class of methods for stiff non-linear initial value problems, Numerical Analysis Dundee, Springer Lecture Notes in Mathematics no. 506, 60-74.

11. _____ (1977), On the relation of G-stability to other concepts for linear multistep methods, Topics in Numerical Analysis III, 67-80, ed. J. H. Miller, Acad. Press, London.

12. _____ (1978), On accuracy and unconditional stability of linear multistep methods for second order differential equations, BIT 18, 133-136.

13. _____ (1978), G-stability is equivalent to A-stability, manuscript submitted to BIT.

14. _____ and O. Nevanlinna (1976), ℓ_2-estimates of the error in the numerical integration of non-linear differential systems, Report TRITA-NA-7607, Computer Science Department, Royal Inst. Technology, Stockholm, Sweden.

15. A. Dinghas (1961), Vorlesungen über Funktionentheorie, Springer-Verlag.

16. W. F. Donoghue, Jr. (1974), Monotone matrix functions and analytic continuation, Springer-Verlag, New York, Heidelberg, Berlin.

17. E. Frank (1946), On the zeros of polynomials with complex coefficients, Bull. Amer. Math. Soc. 52, 144-157.

18. Gantmacher (1959), The theory of matrices, vol. II, Chelsea Publ. Company, New York.

19. C. W. Gear (1971), Numerical initial value problems in ordinary differential equations, Prentice-Hall, Englewood Cliffs, N.J.

20. Y. Genin (1973), A new approach to the synthesis of stiffly stable linear multistep methods, IEEE Trans. on Circuit Theory 20, 352-360.

21. _____ (1974), An algebraic approach to A-stable linear multistep-multiderivative integration formulas, BIT 14, 382-406.

22. _____, Ph. Delsarte and Y. Kamp (1977), Schur parametrization of positive definite Block-Toeplitz systems, M.B.L.E. Res. Lab, Rep. R360.

23. J. M. Gobert, A simple proof of the non A-acceptability of some Padé-approximants of e^{-z}, submitted to Numerische Mathematik.

24. E. A. Guillemin (1949), The mathematics of circuit analysis, The M.I.T. Press, Cambridge, Mass.

25. P. Henrici (1962), Discrete variable methods in ordinary differential equations, Wiley, New York.

26. R. Jeltsch (1978), Stability on the imaginary axis and A-stability of linear multistep methods, BIT 11, 170-174.

27. R. Jeltsch and O. Nevanlinna, personal communication.

28. H.-O. Kreiss (1978), Difference methods for stiff ordinary differential equations, SIAM J. Numer. Anal. 15, 21-58.

29. O. Nevanlinna and W. Liniger (1978), Contractive methods for stiff differential equations, to appear in BIT 1978.

30. F. Odeh and W. Liniger (1977), Nonlinear fixed-h stability of linear multistep formulas, J. Math. Anal. Applic. 61, 691-712.

31. I. Schur (1916), Über Potenzreihen die in Innern des Einheitskreises beschränkt sind, J. Reine Angew. Math. 147, 205-232.

32. J. A. Shohat and J. D. Tamarkin (1963), The problem of moments, AMS Math. Surveys, No. 1, 3rd ed., Providence, R.I.

33. D. D. Šiljak (1971), New algebraic criteria for positive realness, J. Franklin Inst. 291, 109-120.

34. H. S. Wall (1945), Polynomials whose zeros have negative real parts, Amer. Math. Monthly, 52, 308-322.

35. G. Wanner, I. Hairer and S.P. Nørsett (1978), Order stars and stability theorems, to appear in BIT.

36. O. B. Widlund (1967), A note on unconditionally stable linear multistep methods, BIT 7, 65-70.

37. R. Warming and R. Beam (1978), An extension of A-stability to ADI methods. Report at Ames Research Center, NASA, Moffet Field, California.

38. R. Jeltsch (1976), Note on A-stability of multistep multiderivative methods, BIT 16, 74-78.

39. U. Grenander and G. Szegö (1958), Toeplitz forms and their applications. Univ. Calif. Press, Berkeley and Los Angeles.

ACKNOWLEDGMENTS. Funds for the support of this study have been allocated partly by the NASA-Ames Research Center, Moffet Field, California, under interchange No. NCA 2-OR/45-712 and partly by the Department of Energy under Contract no. EY-76-S-03-0326, PA # 30, while the author was a visitor at Stanford University. The author wishes to acknowledge interesting discussions with R. Beam, R. Jeltsch and R. Warming, and express his gratitude to G. Golub for offering excellent working conditions. Thanks are also due to Rosemarie Stampfel for carefully typing the manuscript.

Department of Numerical Analysis
Royal Institute of Technology
S-10044 Stockholm, Sweden

CORRECTION: In order to make sense for $\eta_n = \infty$, the expression for $r_n(z)$ which precedes (4.20) is to be multiplied by $1-i\eta_n$. The formula after (4.21) should be modified accordingly.

Constructive Polynomial Approximation in Sobolev Spaces

Todd Dupont and Ridgway Scott

1. INTRODUCTION.

In this note we give a constructive piecewise polynomial approximation theory having applications to finite element Galerkin methods. The main contribution is a proof of a polynomial approximation lemma popularized by Bramble and Hilbert [1] in which we calculate the constant involved in the error estimate (the proof used in [1] is nonconstructive so that no information is obtained about the constant other than the fact that it is finite). The proof, as well as the result itself, is very closely related to the work of Sobolev [6] on imbedding theorems. Another constructive approach to piecewise polynomial approximation via multipoint Taylor formulae developed by Ciarlet and Wagschall [3] also had an early influence on the present work.

The original objectives of this work were three-fold. First, we simply wanted some idea of the size of the constant in the Bramble-Hilbert lemma. Second, we wanted to know how the constant would change as the underlying domain is perturbed. Finally, we sought a simple proof of the Bramble-Hilbert lemma that could serve as an introduction to polynomial approximation in Sobolev spaces with minimal recourse to complicated functional analysis. While achieving the first

ISBN 0-12-208360-1

two objectives, this note is primarily directed toward the last objective. For this reason we restrict ourselves to approximation by complete polynomials and carry through an application to Lagrange piecewise polynomial interpolation on a domain in $\mathbb{R}^2$ having a curved boundary, including an explicit estimate for the constant in the error estimate. Although we have strived to give reasonable estimates of the constants involved in deriving approximation results, we have made no attempt to get the best possible constants, and indeed the reader will easily see minor improvements (at least) that can result from more precise assumptions on the domains involved.

In a related paper [4], we treat several generalizations of the results presented here. We shall not describe them in detail but shall briefly mention the topics studied. In addition to complete polynomial approximation, [4] also treats approximation by a generalized tensor product polynomial that satisfies a property of commutativity with differentiation similar to the one discussed in the next section. In the tensor product case, the commutativity plays a crucial role in estimating the effect of differentiation on the error, whereas in the complete polynomial case treated here it appears only as a useful tool. A further generalization in [4] in the type of polynomial approximation considered yields, for example, optimal estimates for approximation of harmonic functions by harmonic polynomials. In a different direction, approximation by complete polynomials of functions in fractional order Sobolev spaces is studied, and, finally, a simple observation is used to extend the type of domains for which all of the polynomial approximation results hold. In the present paper, we restrict to domains that are star-shaped with respect to every point in an open ball. However in [4], such results are seen to extend to any domain that is a finite union of domains each of which is star-shaped with respect to a ball separately. Thus domains satisfying a restricted cone property, as in [1], and even more general domains, may be allowed. Furthermore, whereas this paper is concerned primarily with estimates in an L^2 setting, [4] treats the general L^p case. The derivations of error estimates in the cases presented here use more elementary techniques than those required in [4] for the general case.

2. <u>NOTATION</u>.

Let Ω be a bounded open set in $\mathbb{R}^n$ and let dx denote Lebesgue measure in $\mathbb{R}^n$. Denote by $L^p(\Omega)$ the Banach space of measurable functions f such that

$$\|f\|_{L^p(\Omega)} \equiv \left(\int_\Omega |f(x)|^p dx\right)^{1/p} < \infty$$

for $1 \le p < \infty$, with the usual modification when $p = \infty$. Let $\mathbb{N}$ denote the nonnegative integers. For any $\alpha \in \mathbb{N}^n$, let $D^\alpha u = (\partial/\partial x_1)^{\alpha_1} \cdots (\partial/\partial x_n)^{\alpha_n} u$ denote the distributional derivative of $u \in \mathcal{D}'(\Omega)$, the space of Schwartz distributions on Ω. Recall that $|\alpha| \equiv \alpha_1 + \ldots + \alpha_n$ and $\alpha! = \alpha_1! \alpha_2! \ldots \alpha_n!$ for $\alpha \in \mathbb{N}^n$. When u is such that $D^\alpha u$ may be identified with a square integrable function for all $\alpha \in \mathbb{N}^n$ such that $|\alpha| = m$, define

$$|u|_m = |u|_{m,\Omega} = \left(\sum_{|\alpha|=m} \|D^\alpha u\|^2_{L^2(\Omega)}\right)^{1/2} .$$

Similarly define

$$\|u\|_m = \|u\|_{m,\Omega} = \left(\sum_{j=0}^{m} |u|_j^2\right)^{1/2} ,$$

and let $H^m(\Omega)$ denote the Hilbert space having $\|\cdot\|_m$ as norm. Let $C^\infty(\Omega)$ denote the set of infinitely differentiable functions on Ω and $C_0^\infty(\Omega)$ its subspace consisting of functions having compact support in Ω.

Given a nonnegative integer m, let $\mathcal{P}_m$ denote the space of polynomials of degree less than m in n variables.

A subset Ω of $\mathbb{R}^n$ is said to be star-shaped with respect to the point x if $y \in \Omega$ implies $x + \theta(y-x) \in \Omega$ for all $\theta \in [0,1]$.

Given a finite set S, let #S denote the number of elements in S.

Let ω_n denote the measure of the unit sphere in $\mathbb{R}^n$ (i.e., the unit (n-1)-sphere).

3. SOBOLEV REPRESENTATION.

We begin by recalling a constructive polynomial projection operator used by Sobolev [6] in proving imbedding theorems. Let B be an open ball in $\mathbb{R}^n$, and let $\varphi \in C_0^\infty(B)$ be such that $\int_B \varphi(x)dx = 1$. For m any positive integer and $u \in C^\infty(B)$, define

$$Q^m u(x) = \int_B \varphi(y) \sum_{|\alpha|<m} D^\alpha u(y) \frac{(x-y)^\alpha}{\alpha!} dy. \tag{3.1}$$

Note that this is simply the Taylor polynomial approximation

$$T^m u(x) = \sum_{|\alpha|<m} D^\alpha u(y) \frac{(x-y)^\alpha}{\alpha!} \tag{3.2}$$

averaged over B with φ as a weighting function and that Q^m is a projection onto the space of polynomials of degree less than m, $\mathcal{P}_m$. Since the Taylor polynomial T^m satisfies the commutativity relation $D^\alpha T^m u = T^{m-|\alpha|} D^\alpha u$ for $|\alpha| < m$ (as is easily proved by induction), the same relation follows for Q^m by simply differentiating under the integral, namely,

$$D^\alpha Q^m u = Q^{m-|\alpha|} D^\alpha u \tag{3.3}$$

for any $u \in C^\infty(B)$ and $|\alpha| < m$.

Integrating by parts in (3.1), we find that Q^m may also be written as

$$Q^m u(x) = \int_B u(y) \Big(\sum_{|\alpha|<m} (-1)^{|\alpha|} D_y^\alpha [\varphi(y)(x-y)^\alpha/\alpha!] \Big) dy$$
$$\equiv \sum_{|\alpha|<m} x^\alpha \int_B u(y)\, \psi_\alpha(y)\, dy, \tag{3.4}$$

where $\psi_\alpha \in C_0^\infty(B)$. Thus Q^m may be extended to be defined on $\mathcal{D}'(B)$, and since differentiation is a continuous operation on $\mathcal{D}'$, (3.3) is seen to hold for all $u \in \mathcal{D}'(B)$. It also follows easily from (3.4) that Q^m is a Lipschitz continuous mapping of $L^1(B)$ into the Banach space $\mathcal{P}_m$.

Of interest in the sequel will be the error $u - Q^m u$. Since $Q^m u$ is a polynomial and hence defined on all of $\mathbb{R}^n$, it is natural to ask how well $Q^m u$ agrees with u on any domain containing B where u may be defined. Let Ω be a bounded open set in $\mathbb{R}^n$ that is star-shaped with respect to every point in B. For any positive integer m and $u \in C^\infty(\Omega)$, define

$$R^m u(x) = m \sum_{|\alpha|=m} \int_\Omega D^\alpha u(x) k_\alpha(x,y)\, dy, \tag{3.5}$$

where

$$k_\alpha(x,y) \equiv (1/\alpha!)\ (x-y)^\alpha\ k(x,y)$$

and

$$k(x,y) \equiv \int_0^1 s^{-n-1}\ \varphi(x+s^{-1}(y-x))\, ds. \tag{3.6}$$

Notice that the support of $k(x,\cdot)$, and hence the region of integration in (3.6), is contained in the convex hull of $\{x\} \cup B$, which is compactly contained in Ω. Further observe that, if $d \equiv \operatorname{diam}(\Omega)$,

$$|k(x,y)| = \Big| \int_0^1 s^{-n-1}\ \varphi(x+s^{-1}(y-x))\, ds \Big|$$
$$= \Big| \int_{|x-y|/d}^1 s^{-n-1}\ \varphi(x+s^{-1}(y-x))\, ds \Big|$$
$$\le \gamma |x-y|^{-n}$$

for all $x,y \in \Omega$, where $\gamma \equiv (\sup_{x\in B} \varphi(x))\, d^n/n$. Thus each $k_\alpha(x,\cdot)$ is a well defined $L^1(\Omega)$ function if $|\alpha| > 0$. Therefore R^m is indeed defined on $C^\infty(\Omega)$.

THEOREM 3.1. *Let Ω be a bounded open set in $\mathbb{R}^n$ that is star-shaped with respect to every point in an open ball B. Let $\varphi \in C_0^\infty(B)$ be such that $\int_B \varphi(x)dx = 1$, and let Q^m and R^m be the associated operators defined by (3.1) and (3.5) respectively. Then for any positive integer m and any $u \in C^\infty(\Omega)$,*

$$u(x) = Q^m u(x) + R^m u(x)$$

for all $x \in \Omega$.

Remark. The proof of this theorem may be found, eg., in Sobolev's book [6]. More recently, related representation theorems used in the spirit of Sobolev's original work can be found in [2]. For completeness, we include its simple derivation here.

Proof of Theorem 3.1. For $x \in \Omega$ and $y \in B$, Taylor's theorem with integral remainder implies that

$$u(x) = \sum_{|\alpha|<m} \frac{(x-y)^\alpha}{\alpha!} D^\alpha u(y)$$

$$+ m \sum_{|\alpha|=m} \frac{(x-y)^\alpha}{\alpha!} \int_0^1 s^{m-1} D^\alpha u(x+s(y-x))ds.$$

Multiply by $\varphi(y)$ and integrate with respect to y to obtain

$$u(x) = Q^m u(x) + m \sum_{|\alpha|=m} \frac{1}{\alpha!} \int_B \varphi(y)(x-y)^\alpha \int_0^1 s^{m-1} D^\alpha u(x+s(y-x))dsdy.$$

Now use the change of variables $z = x + s(y-x)$ and Fubini's theorem to see that

$$\int \varphi(y)(x-y)^\alpha \int_0^1 s^{m-1} D^\alpha u(x+s(y-x))dsdy$$

$$= \int_0^1 \int \varphi(y)(x-y)^\alpha s^{m-1} D^\alpha u(x+s(y-x))dyds$$

$$= \int_0^1 \int \varphi(x+s^{-1}(z-x))(x-z)^\alpha s^{-1} D^\alpha u(z)\ s^{-n}\ dzds$$

$$= \int (x-z)^\alpha D^\alpha u(z)\ [\int_0^1 \varphi(x+s^{-1}(z-x))\ s^{-n-1}\ ds]\ dz$$

$$= \alpha! \int D^\alpha u(z)\ k_\alpha(x,z)\ dz. \ //$$

4. THE BRAMBLE-HILBERT LEMMA.

THEOREM 4.1. *Let* Ω *be a bounded open set in* $\mathbb{R}^n$ *of diameter* d *that is star-shaped with respect to every point in an open ball* B *of diameter* ρd. *Let* m *be a positive integer and* $u \in H^m(\Omega)$. *Then*

$$\inf_{P \in \mathcal{P}_m} \left(\sum_{j=0}^{m} d^{2j} |u-P|_j^2 \right)^{1/2} \leq C_1 d^m |u|_m,$$

where

$$C_1 = (1+(2/\rho)^{2n} \nu_{m,n}^2)^{1/2}$$

and

$$\nu_{m,n} \equiv \left(\sum_{j=0}^{m-1} \#\{\alpha \in \mathbb{N}^n : |\alpha|=j\} \sum_{|\beta|=m-j} (\beta!)^{-2} \right)^{1/2}.$$

Proof. Because of density, it suffices to assume that $u \in C^\infty(\Omega) \cap H^m(\Omega)$. Let $\varphi \in C_0^\infty(B)$ be such that $\int_B \varphi(x)\ dx = 1$ and let Q^m be the associated polynomial projector constructed in the previous section. Since $|Q^m u|_m = 0$, it suffices to show that

$$\left(\sum_{j=0}^{m-1} d^{2j} |u-Q^m u|_j^2 \right)^{1/2} \leq c\ d^m |u|_m \qquad (4.1)$$

with $c = (2/\rho)^n \nu_{m,n}$. By the commutativity result (3.3) and Theorem 3.1,

$$|u-Q^m u|_j^2 = \sum_{|\alpha|=j} \|D^\alpha (u-Q^m u)\|_0^2$$

$$= \sum_{|\alpha|=j} \|D^\alpha u - Q^{m-j} D^\alpha u\|_0^2$$

$$= \sum_{|\alpha|=j} \|R^{m-j} D^\alpha u\|_0^2$$

Now for each α such that $|\alpha| = j$,

$$\|R^{m-j}D^{\alpha}u\|_0^2 = \int_{\Omega} ((m-j) \sum_{|\beta|=m-j} \int_{\Omega} D^{\alpha+\beta}u(y)k_{\beta}(x,y)dy)^2 \, dx$$

$$\le \int_{\Omega} ((m-j) \sum_{|\beta|=m-j} \frac{1}{\beta!} \int_{\Omega} |D^{\alpha+\beta}u(y)| \; \gamma \; |x-y|^{m-j-n} \, dy)^2 \, dx$$

$$\le \gamma^2(m-j)^2 \; (\sum_{|\beta|=m-j} (\beta!)^{-2})$$

$$\times \sum_{|\beta|=m} \int_{\Omega} (\int_{\Omega} |D^{\beta}u \qquad |x-y|^{m-j-n} \, dy)^2 \, dx$$

$$\le \gamma^2(m-j)^2 \; (\sum_{|\beta|=m-j} (\beta!)^{-2})$$

$$\times \sum_{|\beta|=m} \int_{\Omega} (\int_{\Omega} |D^{\beta}u(y)|^2 \; |x-y|^{m-j-n} \, dy) (\int_{\Omega} |x-y|^{m-j-n} dy) \, dx,$$

where we used Schwarz's inequality at the last step. Note that

$$\int_{\Omega} |x-y|^{m-j-n} \, dy \le \omega_n \int_0^d r^{m-j-1} \, dr$$
$$= \omega_n \; d^{m-j}/(m-j), \qquad (4.2)$$

where ω_n is the measure of the unit (n-1)-sphere. Thus

$$\|R^{m-j}D^{\alpha}u\|_0^2 \le \gamma^2(m-j) \; (\sum_{|\beta|=m-j} (\beta!)^{-2}) \; \omega_n d^{m-j}$$

$$\times \sum_{|\beta|=m} \int_{\Omega}(\int_{\Omega} |D^{\beta}u(y)|^2 \; |x-y|^{m-j-n} \, dy) \, dx.$$

Using Fubini's theorem and (4.2) with the roles of x and y reversed, we obtain

$$\|R^{m-j}D^{\alpha}u\|_0^2 \le \gamma^2(m-j) \; (\sum_{|\beta|=m-j} (\beta!)^{-2}) \; \omega_n \; d^{m-j}$$

$$\times \sum_{|\beta|=m} \int_{\Omega} |D^{\alpha}u(y)|^2 \; (\int_{\Omega} |x-y|^{m-j-n} \, dx) \, dy$$

$$\le \gamma^2(\sum_{|\beta|=m-j} (\beta!)^{-2}) \; \omega_n^2 \; d^{2(m-j)} \; |u|_m^2.$$

Summing appropriately, we obtain (4.1) with $c = \gamma \; \omega_n \nu_{m,n}$. Recall that φ was arbitrary, except that $\varphi \in C_0^{\infty}(B)$ and $\int_B \varphi(x) \, dx = 1$, and that $\gamma = (\sup_{x \in B} |\varphi(x)|) \; d^n/n$. We can choose

φ arbitrarily close to a constant multiple of the characteristic function of B, and hence $\sup_{x\in B}|\varphi(x)|$ may be chosen arbitrarily close to 1/meas B = $(\frac{1}{n}(\frac{1}{2}\rho d)^n\ \omega_n)^{-1} = n\ 2^n(\rho d)^{-n}\ \omega_n^{-1}$. Thus γ may be chosen to be $(2/\rho)^n\ \omega_n^{-1}$. //

Remark. What we have proved is in fact sharper, namely that

$$\inf_{P\in\mathcal{P}_m}\ \max_{0\le j\le m} \frac{1}{\mu_j}\, d^j\,|u-P|_j \le (2/\rho)^n\ d^m |u|_m ,$$

where

$$\mu_j = (\#\{\alpha\in\mathbb{N}^n : |\alpha|=j\} \sum_{|\beta|=m-j} (\beta!)^{-2})^{1/2}$$

for $j = 0,1,\ldots,m-1$, and $\mu_m = (\rho/2)^n$. When j is small and m is large, the constant μ_j is quite small. For example, suppose n = 2. Then

$$\sum_{|\beta|=k} (\beta!)^{-2} = \sum_{i=0}^{k} (i!(k-i)!)^{-2}$$
$$= (k!)^{-2} \sum_{i=0}^{k} \binom{k}{i}^2 .$$

Using Hölder's inequality, we find that

$$\sum_{i=0}^{k} \binom{k}{i}^2 \le \left[\max_i \binom{k}{i}\right] \sum_{i=0}^{k} \binom{k}{i} \le 2^{2k-1}.$$

Therefore

$$\sum_{|\beta|=k} (\beta!)^{-2} \le 2^{2k-1}/(k!)^2. \tag{4.3}$$

Since $\#\{\alpha\in\mathbb{N}^2 : |\alpha|=j\} = j + 1$, we find

$$\mu_j \le \sqrt{(j+1)/2}\ \ 2^{m-j}/(m-j)!\,.$$

Since $\nu_{m,2} = (\sum_{j=0}^{m-1} \mu_j^2)^{1/2}$, we also find that

$$\nu_{m,2}^2 \le \sum_{i=1}^{m} (m-i+1)\ 2^{2i-1}/(i!)^2$$
$$\le m \sum_{i=1}^{m} 2^i/i!$$
$$\le m\ e^2.$$

Therefore, $\nu_{m,2} \le \sqrt{m}\, e$. //

The following theorem is closely related to the previous one and in fact follows from it via Sobolev's inequality. It will be central to the estimation of interpolation error given in the next section, and hence we give an independent proof.

THEOREM 4.2. *Let* Ω *be a bounded open set in* $\mathbb{R}^n$ *of diameter* d *that is star-shaped with respect to every point of an open ball* B *of diameter* ρd. *Let* k *be a nonnegative integer and let* m *be an integer such that* $m > k + n/2$. *Then for all* $u \in H^m(\Omega)$ *and* $\alpha \in \mathbb{N}^n$ *such that* $|\alpha| \le k$,

$$\inf_{P \in \mathcal{P}_m} \max_{|\alpha| \le k} d^{|\alpha|} \|D^\alpha (u-P)\|_{L^\infty(\Omega)} \le C_2\, d^{m-n/2} |u|_m ,$$

where

$$C_2 = m\, \omega_n^{-1/2} (2(m-k)-n)^{-1/2} \max_{0 \le j \le k} \Big(\sum_{|\beta|=m-j} (\beta!)^{-2} \Big)^{1/2} (2/\rho)^n .$$

Proof. It suffices to assume that $u \in C^\infty(\Omega) \cap H^m(\Omega)$ and to show that

$$\|D^\alpha (u-Q^m u)\|_{L^\infty(\Omega)} \le C_2'\, d^{m-n/2-|\alpha|} |u|_m$$

for all $|\alpha| \le k$, where C_2' may be made arbitrarily close to C_2 by choosing φ appropriately as in Theorem 4.1. For $x \in \Omega$, we have

$$|D^\alpha (u-Q^m u)(x)| = |(D^\alpha u - Q^{m-|\alpha|} D^\alpha u)(x)|$$

$$= |R^{m-|\alpha|} D^\alpha u(x)|$$

$$= |(m-|\alpha|) \sum_{|\beta|=m-|\alpha|} \frac{1}{\beta!} \int_\Omega D^{\alpha+\beta} u(y)\, k_\beta(x,y)\, dy|$$

$$\le (m-|\alpha|)\, \gamma \sum_{|\beta|=m-|\alpha|} \frac{1}{\beta!} \int_\Omega |D^{\alpha+\beta} u(y)|\, |x-y|^{m-|\alpha|-n}\, dy$$

$$\le (m-|\alpha|)\, \gamma \Big(\sum_{|\beta|=m-|\alpha|} \frac{1}{\beta!} \|D^{\alpha+\beta} u\|_0 \Big) \Big(\int_\Omega |x-y|^{2(m-|\alpha|-n)}\, dy \Big)^{1/2}$$

$$\leq (m-|\alpha|)\ \gamma\Big(\sum_{|\beta|=m-|\alpha|}\frac{1}{\beta!}\|D^{\alpha+\beta}u\|_0\Big)\ \omega_n^{1/2}\Big(\int_0^d r^{2(m-|\alpha|)-n-1}dr\Big)^{1/2}$$

$$\leq (m-|\alpha|)\ \gamma\omega_n^{1/2}\ d^{m-|\alpha|-n/2}(2(m-|\alpha|)-n)^{-1/2}$$

$$\times\ \Big(\sum_{|\beta|=m-|\alpha|}\frac{1}{\beta!}\|D^{\alpha+\beta}u\|_0\Big)$$

$$\leq (m-|\alpha|)\ \gamma\omega_n^{1/2} d^{m-|\alpha|-n/2}(2(m-|\alpha|)-n)^{-1/2}\Big(\sum_{|\beta|=m-|\alpha|}(\beta!)^{-2}\Big)^{1/2}|u|_m.$$

Letting γ tend to the value $(2/\rho)^n\ \omega_n^{-1}$ as in the proof of the previous theorem, the conclusion is reached. //

Remarks. First note that since $Q^m u$ was used to approximate u in the proofs of both Theorem 4.1 and 4.2, we can conclude that the stated approximations hold simultaneously, i.e.,

$$\inf_{P\in\mathcal{P}_m}\Big(\Big(\sum_{j=0}^{m} d^{2j}|u-P|_j^2\Big)^{1/2}+\max_{|\alpha|\leq k} d^{|\alpha|+n/2}\|D^\alpha(u-P)\|_{L^\infty(\Omega)}\Big)$$
$$\leq (C_1+C_2)\ d^m|u|_m.$$

Second, observe that Sobolev's inequality follows almost immediately from the proof of Theorem 4.2. All that remains to be estimated, in view of the triangle inequality, is $\|D^\alpha Q^m u\|_{L^\infty(\Omega)}$. It follows from (3.4) that

$$\|D^\alpha Q^m u\|_{L^\infty(\Omega)} \leq \text{constant}\ \|u\|_{L^1(\Omega)},$$

where we shall <u>not</u> evaluate the constant, but the eager reader may easily do so. Thus we obtain

$$\max_{|\alpha|\leq k}\|D^\alpha u\|_{L^\infty(\Omega)} \leq \text{constant}\ (\|u\|_{L^1(\Omega)}+|u|_m).$$

This is essentially the proof used by Sobolev [6] for the imbedding theorems.

Finally, we direct the reader back to the bound (4.3) for the quantity $\sum_{|\beta|=j}(\beta!)^{-2}$ appearing in the definition of C_2 in the case n = 2. //

5. ESTIMATES FOR LAGRANGE INTERPOLATION.

Let $\mathcal{O}$ be a bounded open set in $\mathbb{R}^2$ with smooth boundary $\partial\mathcal{O}$. Let $\mathfrak{I}$ be a family of triangulations $\mathcal{T}$ of $\mathcal{O}$ consisting of triangles T having straight interior edges but (possibly) one curved edge on $\partial\mathcal{O}$. Define $h(\mathcal{T}) = \max\{\text{diam}(T) : T\epsilon\mathcal{T}\}$ and let b(T) be the diameter of the largest open ball with respect to which T is star-shaped. Suppose that $\mathfrak{I}$ satisfies the following nondegeneracy condition with $\rho > 0$:

$$\min_{T\epsilon\mathcal{T}} b(T) \geq \rho h(\mathcal{T}) \quad \text{for all } \mathcal{T}\epsilon\mathfrak{I}. \tag{5.1}$$

Let $\tau = \{(x,y): x,y\geq 0; x+y\leq 1\}$ and define Lagrange interpolation nodes for polynomials of degree less than m on τ to be the $\frac{1}{2}\,m(m+1)$ points in τ whose coordinates have values i/k for some integer i. For each $\mathcal{T}\epsilon\mathfrak{I}$ and each $T\epsilon\mathcal{T}$, define a reference domain T_R by letting T_R be the image of T under an affine mapping that maps the vertices of T onto the set $\{(0,0),(1,0),(0,1)\}$ (there are six such affine maps -- choose one arbitrarily). If T has straight edges, then T_R (or its closure) equals τ, and we define Lagrange interpolation nodes for T via the inverse affine mapping $T \leftarrow T_R = \tau$. If T has a curved edge, we must modify this. For h_0 sufficiently small, depending on the curvature of $\partial\mathcal{O}$ and ρ, if $h(\mathcal{T}) \leq h_0$ and $T\epsilon\mathcal{T}$, then the interior nodes of τ lie in the interior of T_R. In this case, we define nodes for T by mapping the interior nodes of τ back to T, and for the edges of T we take the k-1 points on each edge that divide the edge into k **arcs** of equal length, plus the vertices of T.

As is well known, given any continuous function u defined on $\mathcal{O}$, there is a unique function $\mathcal{J}_{\mathcal{T}}u\epsilon H^1(\Omega)$ such that for all $T\epsilon\mathcal{T}$, $\mathcal{J}_{\mathcal{T}}u|_T$ is a polynomial of degree less than m that agrees with u at the Lagrange interpolation nodes for T defined above, at least provided $h(\mathcal{T}) \leq h_0$ and h_0 is sufficiently small depending on $\partial\mathcal{O}$ and ρ. Furthermore

$$\sup_{x\epsilon T} |\mathcal{J}_{\mathcal{T}}u(x)| \leq K \sup_{x\epsilon T} |u(x)| \quad \text{for all } T\epsilon\mathcal{T}, \tag{5.2}$$

where K is easily computed. For example, for piecewise quadratic interpolation, $K \le 7$ for $h(T) \le h_0$ and h_0 sufficiently small depending on $\partial\mathcal{O}$ and ρ. To see this, consider the Lagrange interpolation basis functions for τ. Each is bounded by 1, and there are six of them, so if T has only straight sides, we may take $K \le 6$. Since the nodes for a T having a curved side are an $O(h^2)$ perturbation from nodes for a straight-edged triangle, we conclude $K \le 7$ is sufficient for h small.

We can estimate the error $u - \mathcal{J}_T u$ as follows. Let $T \in \mathcal{T}$. Then for any $P \in \mathcal{P}_m$

$$\begin{aligned}\|u-\mathcal{J}_T u\|_{L^2(T)} &\le \|u-P\|_{L^2(T)} + \|\mathcal{J}_T(u-P)\|_{L^2(T)} \\ &\le \sqrt{\text{meas}(T)}\ (\|u-P\|_{L^\infty(T)} + \|\mathcal{J}_T(u-P)\|_{L^\infty(T)}) \\ &\le \sqrt{\text{meas}(T)}\ (1+K)\ \|u-P\|_{L^\infty(T)},\end{aligned}$$

where we have used the fact that $\mathcal{J}_T P = P$, Hölder's inequality, and the bound (5.2). Since P was arbitrary and $\sqrt{\text{meas}(T)} \le \sqrt{\pi/4}\ h(T)$, Theorem 4.2 implies that

$$\|u-\mathcal{J}_T u\|_{L^2(T)} \le Ch(T)^m\ |u|_{m,T}\ ,$$

where

$$C \le (m/\sqrt{m-1})\,\Big(\sum_{|\beta|=m} (\beta!)^{-2}\Big)^{1/2} \rho^{-2}(1+K).$$

Squaring and summing over $T \in \mathcal{T}$, we find

$$\|u-\mathcal{J}_T u\|_0 \le Ch(T)^m\ |u|_m\ ,$$

where the norms refer to the domain $\mathcal{O}$. In the case of piecewise quadratic interpolation (m=3), we thus have

$$\|u-\mathcal{J}_T u\|_0 \le 10\rho^{-2}\ h(T)^3\ |u|_3.$$

Remarks. Estimates similar to those above were derived in [5] using a much more complicated technique and without any estimate for the constant involved. Such estimates can also be derived when $\partial\mathcal{O}$ is piecewise smooth; when $\partial\mathcal{O}$ is polygonal and each $T\in T\in\mathfrak{J}$ has only straight sides, the restriction "$h(T) \le h_0$" may be removed.

REFERENCES

1. J. H. Bramble and S. R. Hilbert (1970), Estimation of linear functionals on Sobolev spaces with applications to Fourier transforms and spline interpolation, *SIAM J. Numer. Anal.*, 7, 112-124.

2. V. I. Burenkov (1974), Sobolev's integral representation and Taylor's formula, *Trudy Mat. Inst. Steklov*, 131, 33-38.

3. P. G. Ciarlet and C. Wagschall (1971), Multipoint Taylor formulas and applications to the finite element method, *Numer. Math.*, 17, 84-100.

4. T. Dupont and R. Scott (1978), Polynomial approximation of functions in Sobolev spaces, to appear.

5. R. Scott (1975), Interpolated boundary conditions in the finite element method, *SIAM J. Numer. Anal.* 12, 404-427.

6. S. L. Sobolev (1963), *Applications of Functional Analysis in Mathematical Physics*, American Mathematical Society, Providence.

The submitted manuscript has been authored under contract EY-76-C-02-0016 with the U. S. Department of Energy.

Department of Mathematics
University of Chicago
Chicago, IL 60637

Applied Mathematics Department
Brookhaven National Laboratory
Upton, NY 11973

Questions of Numerical Condition Related to Polynomials

Walter Gautschi

1. INTRODUCTION.

Polynomials (in one variable) permeate much of classical numerical analysis, either in the role of approximators, or as gauge functions for a variety of numerical methods, or in the role of characteristic polynomials of one kind or another. It seems appropriate, therefore, to study some of their basic properties as they relate to computation. In the following we wish to consider one particular aspect of polynomials, namely the extent to which they, or quantities related to them, are sensitive to small perturbations. In other words, we are interested in the numerical condition of polynomials. We shall examine from this angle three particular problem areas: (1) The representation of polynomials (polynomial bases); (2) Algebraic equations; (3) The problem of orthogonalization. Before embarking on these topics, however, we must briefly consider ways of measuring the condition of problems. We do this in the framework of maps from one normed space into another, for which we define appropriate condition numbers.

2. THE CONDITION OF MAPS.

2.1. Nonlinear maps. Let X, Y be normed linear spaces, and let $y = f(x)$ define a map $M: \mathcal{D} \subset X \to Y$, with $\mathcal{D}$ an open domain. Let $\mathring{x} \in \mathcal{D}$ be fixed, and $\mathring{y} = f(\mathring{x})$, and assume that neither $\mathring{x}$ nor $\mathring{y}$ is the zero element in the respective

ISBN 0-12-208360-1

space. The sensitivity of the map M at $\overset{\circ}{x}$, with respect to small relative changes in $\overset{\circ}{x}$, will be measured by the (asymptotic) condition number (cf. Rice (1966))

$$\operatorname{cond}(M;\overset{\circ}{x}) = \lim_{\delta\to 0} \sup_{\|h\|=\delta} \left\{ \frac{\|f(\overset{\circ}{x}+h)-f(\overset{\circ}{x})\|}{\|f(\overset{\circ}{x})\|} \Big/ \frac{\|h\|}{\|\overset{\circ}{x}\|} \right\}, \tag{2.1}$$

provided the limit exists. The number in (2.1) measures the maximum amount by which a relative perturbation of $\overset{\circ}{x}$ (given by $\delta/\|\overset{\circ}{x}\|$) is magnified under the map M, in the limit of infinitesimal perturbations. Maps with large condition numbers are called ill-conditioned.

If M has a Fréchet derivative $[\partial f/\partial x]_0$ at $\overset{\circ}{x}$, then

$$\operatorname{cond}(M;\overset{\circ}{x}) = \frac{\|\overset{\circ}{x}\|}{\|\overset{\circ}{y}\|}\left\|\left[\frac{\partial f}{\partial x}\right]_0\right\| \qquad (\overset{\circ}{y} = f(\overset{\circ}{x})) . \tag{2.2}$$

In the important case of finite-dimensional spaces, $X = \mathbb{R}^n$, $Y = \mathbb{R}^m$, the Fréchet derivative, as is well-known, is the linear map defined by the Jacobian matrix of f. We may then use in (2.2) any family of vector norms and subordinate family of matrix norms (cf. Stewart (1973), p. 177).

For composite maps $K \circ M$, the chain rule for Fréchet derivatives (cf. Ortega & Rheinboldt (1970), p. 62) can be used to show that

$$\operatorname{cond}(K \circ M;\overset{\circ}{x}) \le \operatorname{cond}(K;\overset{\circ}{y})\operatorname{cond}(M;\overset{\circ}{x}). \tag{2.3}$$

If the composite map is known to be ill-conditioned, the inequality (2.3) permits us to infer the ill-conditioning of (at least) one of the component maps.

2.2. Linear maps. If $M : y = f(x)$ is a linear (bounded) map, then

$$\sup_{\|h\|=\delta} \frac{\|f(\overset{\circ}{x}+h) - f(\overset{\circ}{x})\|}{\|h\|} = \sup_{\|h\|=\delta} \frac{\|f(h)\|}{\|h\|}$$

is independent of $\overset{\circ}{x}$ and δ, and equal to the norm of M. Equation (2.1) then reduces to

$$\operatorname{cond}(M;\overset{\circ}{x}) = \frac{\|\overset{\circ}{x}\|}{\|\overset{\circ}{y}\|}\|M\| \qquad (M \text{ linear, } \overset{\circ}{y} = M\overset{\circ}{x}). \tag{2.4}$$

If in addition M is invertible, we can ask for the supremum of (2.4) as $\overset{\circ}{x}$ varies in X (or, equivalently, $\overset{\circ}{y}$ varies in MX), and we find, since $\overset{\circ}{x} = M^{-1}\overset{\circ}{y}$, that

$$\sup_{x \in X} \text{cond}(M;x) = \|M^{-1}\| \, \|M\|. \tag{2.5}$$

The number on the right, usually referred to as the <u>condition number of</u> M, will be denoted by

$$\text{cond } M = \|M^{-1}\| \, \|M\| . \tag{2.6}$$

We have, alternatively,

$$\text{cond } M = \frac{\sup_{x \in X} (\|Mx\| / \|x\|)}{\inf_{x \in X} (\|Mx\| / \|x\|)} . \tag{2.7}$$

Condition numbers, such as those proposed, cannot be expected to do more than convey general guidelines as to the susceptibility of the respective maps to small changes in their domains. By their very definition they reflect "worst case" situations and therefore are inherently conservative measures.

3. THE CONDITION OF POLYNOMIAL BASES.

Let $\mathbb{P}_{n-1}$ denote the class of (real) polynomials of degree $\leq n-1$, and let $p_1, p_2, \ldots, p_n$ be a basis in $\mathbb{P}_{n-1}$. For any $p \in \mathbb{P}_{n-1}$, we denote by $u_1, u_2, \ldots, u_n$ the coefficients of p with respect to this basis,

$$p(x) = \sum_{k=1}^{n} u_k \, p_k(x) . \tag{3.1}$$

We wish to determine how strongly the values of p on some given finite interval $[a,b]$ react to small perturbations in the coefficients u_k, and vice versa, how the coefficients of p are affected by small changes in p.

The question may be formalized as one concerning the condition of the linear map $M_n : \mathbb{R}^n \to \mathbb{P}_{n-1}[a,b]$, which associates to each vector $u^T = [u_1, u_2, \ldots, u_n] \in \mathbb{R}^n$ the polynomial p in (3.1), restricted to $[a,b]$,

$$(M_n u)(x) = \sum_{k=1}^{n} u_k \, p_k(x) , \qquad a \leq x \leq b. \tag{3.2}$$

We shall use the notation u_p to denote the coefficient vector of p,

$$u_p = M_n^{-1} p , \quad p \in \mathbb{P}_{n-1} . \tag{3.3}$$

We are thus interested in the condition number of M_n, cf. (2.6),

$$\text{cond } M_n = \|M_n^{-1}\| \, \|M_n\|, \tag{3.4}$$

in particular, how fast it grows as $n \to \infty$, and how this growth depends on the particular interval chosen.

For definiteness, we consider only uniform norms, i.e., $\|u\| = \max_{1\le k\le n} |u_k|$ in $\mathbb{R}^n$, and $\|p\| = \max_{a\le x\le b} |p(x)|$ in $\mathbb{P}_{n-1}[a,b]$.

3.1. Power basis. For the power basis

$$p_k(x) = x^{k-1}, \qquad k = 1,2,\ldots,n, \tag{3.5}$$

it is natural to assume an interval $[a,b]$ that contains the origin. We shall do so in the following, although other intervals could also be treated (in fact more easily). For definiteness, we assume further that $[a,b]$ is centered to the right of the origin, i.e., $0 \le |a| \le b$. It then follows immediately that

$$\|M_n\| = \sup_{\|u\|=1} \max_{a\le x\le b} \left|\sum_{k=1}^{n} u_k x^{k-1}\right| = \sum_{k=1}^{n} b^{k-1},$$

hence

$$\|M_n\| = \frac{b^n-1}{b-1}. \tag{3.6}$$

(It is understood, here and below, that the value of the function on the right equals n if $b=1$.)

For the inverse map M_n^{-1} we have

$$\|M_n^{-1}\| = \sup_{\|p\|=1} \max_{1\le k\le n} \frac{|p^{(k-1)}(0)|}{(k-1)!} = \max_{1\le k\le n} \sup_{\|p\|=1} \frac{|p^{(k-1)}(0)|}{(k-1)!}.$$

Therefore, in terms of the linear functionals $\lambda_k : \mathbb{P}_{n-1}[a,b] \to \mathbb{R}$ defined by $\lambda_k p = p^{(k-1)}(0)/(k-1)!$,

$$\|M_n^{-1}\| = \max_{1\le k\le n} \|\lambda_k\|. \tag{3.7}$$

Our problem thus reduces to determining the norm of λ_k. This is related to the problem of best uniform approximation of binomials $f_{n,\tau}(x) = (1-|\tau|)x^n + \tau x^{n-1}$, where $-1 \le \tau \le 1$, by polynomials g of degree $\le n-2$, which in turn gives rise to the Zolotarev polynomials

$$z_{n,\tau}(x) = \frac{1}{E_{n,\tau}}(f_{n,\tau}(x) - g^*_{n,\tau}(x)), \quad -1 \le \tau \le 1,$$

where $E_{n,\tau} = \inf_{g \in \mathbb{P}_{n-2}} \|f_{n,\tau} - g\| = \|f_{n,\tau} - g^*_{n,\tau}\|$. The extremal for the functional λ_k, indeed, is a Zolotarev polynomial (of degree $n-1$, since we are working with $\mathbb{P}_{n-1}[a,b]$), that is, for $2 \le k \le n$,

$$\|\lambda_k\| = \sup_{\|p\|=1} |\lambda_k p| = |\lambda_k z_{n-1,\tau}| \text{ for some } \tau \in [-1,1] \qquad (3.8)$$

(see, e.g., Schönhage (1971, Satz 6.11)). Unfortunately, if the interval $[a,b]$ is arbitrary, the value of the parameter τ in (3.8) is not easily expressible, and may be different for different values of k. The exact determination of $\|M_n^{-1}\|$ in (3.7) is quite cumbersome (cf. Voronovskaja (1970, Ch. III and the appendix by V. A. Gusev)). Upper bounds for $\|M_n^{-1}\|$ are obtained in Gautschi (1979a, Theorem 4.1).

For the power basis (3.5), the most natural interval, however, is an interval symmetric about the origin, $[-\omega,\omega]$, $\omega > 0$. In this case, the Zolotarev polynomials reduce to Chebyshev polynomials of the first kind (Schönhage (1971), p. 167). Making use of this, it then follows readily from (3.6) and (3.7) that

$$\text{cond } M_n = \frac{\omega^n - 1}{\omega - 1} \max\{\|u_{T_{n-1}(x/\omega)}\|, \|u_{T_{n-2}(x/\omega)}\|\}, \qquad (3.9)$$

where T_m denotes the Chebyshev polynomial of degree m, and $u_{T_m(x/\omega)}$ the coefficient vector of $T_m(x/\omega)$; cf. (3.3).

Using asymptotic estimates of $\|u_{T_m(x/\omega)}\|$ as $m \to \infty$, Gautschi (1979a, Eq. (2.2)), it can be deduced from (3.9) that

$$(\text{cond } M_n)^{\frac{1}{n}} \sim \begin{cases} 1 + \sqrt{1+\omega^2}, & \omega \ge 1, \\ \dfrac{1 + \sqrt{1+\omega^2}}{\omega}, & \omega < 1, \end{cases} \quad \text{as } n \to \infty. \qquad (3.10)$$

The condition of M_n on $[-\omega,\omega]$ thus grows exponentially with n, the asymptotic growth rate being smallest (equal to $1+\sqrt{2}$) when $\omega = 1$. Some numerical values are shown in Table 3.1.[†]

n	$(\text{cond } M_n)^{1/n}$				
	$\omega = .1$	$\omega = .2$	$\omega = 1$	$\omega = 5$	$\omega = 10$
5	9.767	5.743	2.091	3.789	6.444
10	13.977	7.579	2.377	4.616	8.027
20	16.719	8.706	2.437	5.210	9.252
40	18.286	9.330	2.447	5.588	10.023
⋮	⋮	⋮	⋮	⋮	⋮
∞	20.050	10.099	2.414	6.099	11.050

Table 3.1. The condition of M_n on $[-\omega,\omega]$

Similar results hold for intervals $[0,\omega]$, $\omega > 0$, in which case (Gautschi (1979a))

$$(\text{cond } M_n)^{\frac{1}{n}} \sim \begin{cases} 2+\omega + 2\sqrt{1+\omega}\,, & \omega \geq 1, \\ \dfrac{2+\omega + 2\sqrt{1+\omega}}{\omega}, & \omega < 1, \end{cases} \quad \text{as } n \to \infty . \tag{3.11}$$

Again, the minimum growth rate $(= (1+\sqrt{2})^2)$ occurs when $\omega = 1$.

[†]The information in Table 3.1 might suggest that the asymptotic growth rate is approached monotonically. The reader, however, will have noticed that the limit rate in the case $\omega = 1$ is smaller than the seemingly increasing approach rates! In reality, the approach is indeed monotone, if $\omega < 1$, but changes from increasing to decreasing, if $\omega = 1$, and from decreasing to increasing, if $\omega > 1$. The changeover occurs near $n = 35$, if $\omega = 1$ (hence is not visible in Table 3.1), and near $(e/2\pi)\omega$, if $\omega \gg 1$. The latter would begin to be visible in Table 3.1 if $(e/2\pi)\omega \doteq 10$, i.e., $\omega \doteq 23$. The reason for this behavior can be found in the more precise relations $\text{cond } M_n \sim (\gamma^2 n)^{1/2} \rho^n$ if $\omega = 1$, and $\text{cond } M_n \sim (\gamma^2/n)^{1/2} \rho^n$, if $\omega \neq 1$, where ρ is the limit rate and $\gamma=\gamma(\omega)$ can be explicitly computed; Gautschi (1979a).

It is interesting to note that exponential growth of the condition is also observed for piecewise polynomial functions, if represented in terms of normalized B-splines. In fact, for splines of degree $k-1$, the condition of the B-spline basis is known to lie between $(1 - 1/k)2^{k-3/2}$ and $2k \cdot 9^k$; cf. de Boor (1972), Lyche (1978). Empirical evidence seems to suggest that the condition is indeed $O(2^k)$; de Boor (1976).

3.2. Bases of orthogonal polynomials. We now consider the case of an orthogonal basis, i.e.,

$$p_k(x) = \pi_{k-1}(x), \qquad k = 1,2,\ldots,n, \tag{3.12}$$

where $\pi_0, \pi_1, \ldots, \pi_{n-1}$ are the first n of a sequence of polynomials orthogonal on the (finite) interval $[a,b]$ with respect to a non-negative measure $d\sigma(x)$. We consider the condition of this basis on the interval of orthogonality, $[a,b]$. Since the coefficients u_k in (3.1) of any polynomial $p \in \mathbb{P}_{n-1}$ are now representable as Fourier coefficients of p, it is easy to estimate the condition of M_n with the aid of Schwarz' inequality. One finds (Gautschi (1972))

$$\text{cond } M_n \le \max_{1\le k\le n} \left(\frac{\mu_0}{h_{k-1}}\right)^{1/2} \cdot \max_{a\le x\le b} \sum_{k=1}^{n} |\pi_{k-1}(x)|, \tag{3.13}$$

where

$$\mu_0 = \int_a^b d\sigma(x), \quad h_k = \int_a^b \pi_k^2(x)\,d\sigma(x), \quad k = 0,1,\ldots . \tag{3.14}$$

The first maximum in (3.13) is a bound for $\| M_n^{-1} \|$, the second an obvious bound for $\| M_n \|$. It should be noted that neither cond M_n nor the bound in (3.13) is invariant under different normalizations of the orthogonal polynomials $\{\pi_{k-1}\}$, and the bound indeed is minimized in the case of an orthonormal system.

It follows from (3.13) that the condition of an orthogonal basis, typically, exhibits only polynomial growth in n. For Chebyshev polynomials $\pi_r = T_r$ on $[-1,1]$, for example, one finds

$$\text{cond } M_n \le 2^{1/2} n \quad (\pi_r = T_r),$$

while for Legendre polynomials $\pi_r = P_r$ on $[-1,1]$,

$$\text{cond } M_n \le n(2n-1)^{1/2} \quad (\pi_r = P_r).$$

The improvement over the power basis is substantial.

3.3. Lagrangian bases. All bases $\{p_k\}$ considered previously have the property that $\deg p_k = k-1$, $k = 1,2,3,\ldots$. We now consider an example of a basis in which each p_k is a polynomial of degree $n-1$, namely the familiar Lagrange polynomials

$$p_k(x) = \ell_k(x), \qquad \ell_k(x) = \prod_{\substack{\nu=1\\ \nu\neq k}}^{n} \frac{x-s_\nu}{s_k-s_\nu}, \qquad k = 1,2,\ldots,n ,$$

corresponding to a set of distinct nodes $s_1,s_2,\ldots,s_n$ in $[a,b]$. Lagrange's interpolation formula

$$p(x) = \sum_{k=1}^{n} p(s_k)\, \ell_k(x)$$

shows immediately that $u_k = p(s_k)$ in (3.1). By standard arguments in approximation theory, one finds $\|M_n\| = L_n$, $\|M_n^{-1}\| = 1$, where

$$L_n = \max_{a\leq x\leq b} \sum_{k=1}^{n} |\ell_k(x)|$$

is the Lebesgue constant for the nodes s_ν. Consequently (cf. also de Boor (1978b)),

$$\operatorname{cond} M_n = L_n . \tag{3.15}$$

By a result of Faber and Bernstein, Natanson (1965, p. 24), one has

$$L_n > \frac{\ln n}{8\sqrt{\pi}}$$

for arbitrary (distinct) nodes s_ν, while for Chebyshev nodes, on the other hand,

$$L_n \sim \frac{2}{\pi} \ln n , \quad n \to \infty$$

(see, e.g., Rivlin (1974, p. 18)). The basis consisting of the Lagrange polynomials $\{\ell_k\}$ for Chebyshev nodes, therefore, is optimally conditioned among all Lagrangian bases, and indeed among all polynomial bases (de Boor (1978a)), in the sense of attaining the optimal growth rate $O(\ln n)$.

4. THE CONDITION OF ALGEBRAIC EQUATIONS.

We now turn our attention to roots of algebraic equations and their sensitivity to small changes in the coefficients. (We assume that the equation is expressed linearly in terms of basis polynomials.) An interesting, though largely unexplored, aspect of this question is the manner in which this sensitivity depends on the choice of polynomial basis. By far best understood is the case of equations expressed in the usual power form.

In order to give a formal statement of the problem, we assume, first of all, that the basis polynomials p_k have $\deg p_k = k-1$, $k = 1,2,\ldots$, so that an algebraic equation of exact degree n can be written in normalized form

$$p(x) = 0 , \qquad p(x) = p_{n+1}(x) + \sum_{k=1}^{n} u_k p_k(x) , \tag{4.1}$$

with leading coefficient 1. (To enhance clarity, we sometimes write $p(u;x)$ instead of $p(x)$, where $u = [u_1,u_2,\ldots,u_n]^T$.)

In general, one might be interested in just one, or in several, or collectively in all the roots of the equation, and again, there may be one single coefficient, or several, or all of them that are subject to perturbation. We treat all these cases in one, by considering q (simple) roots $\underline{\mathring{\xi}} = [\mathring{\xi}_1,\mathring{\xi}_2,\ldots,\mathring{\xi}_q]^T$, $1 \le q \le n$, of (4.1), corresponding to $u=\mathring{u}$, and by introducing a multi-index $\underline{k} = (k_1,k_2,\ldots,k_p)$, $1 \le k_1 < k_2 < \ldots < k_p \le n$, to indicate which of the coefficients $\mathring{u}_k$ are to undergo changes. We write $\underline{k}^c$ for the multi-index complementary to $\underline{k}$, and denote by $\underline{u} \in \mathbb{R}^n$ the vector whose k-th component is $\mathring{u}_k$, if $k \in \underline{k}^c$, and u_k, if $k \in \underline{k}$. There will be a neighborhood $\mathcal{D} = N(\mathring{u}_{\underline{k}}) \subset \mathbb{R}^p$ such that the equation (4.1) with $u = \underline{u}$, $u_{\underline{k}} \in \mathcal{D}$, continues to have q simple zeros $\underline{\xi} = [\xi_1,\xi_2,\ldots,\xi_q]^T$ and $\underline{\xi} \to \underline{\mathring{\xi}}$ as $\underline{u} \to \mathring{u}$. We assume that neither $\underline{\mathring{\xi}}$, nor $\mathring{u}_{\underline{k}}$, is the zero vector in the space $\mathbb{C}^q$ and $\mathbb{R}^p$, respectively. (Clearly, $\underline{\mathring{\xi}} \neq \underline{0}$ if $q > 1$, since each $\mathring{\xi}_j$ is simple.) Our interest, then, is in the condition of the map $M_{\underline{k},q} : \mathcal{D} \subset \mathbb{R}^p \to \mathbb{C}^q$ defined by

$$M_{\underline{k},q} : \underline{\xi} = \underline{f}(u_{\underline{k}}), \quad u_{\underline{k}} \in \mathcal{D} \subset \mathbb{R}^p ,$$

where $p(\underline{u};f_j(u_{\underline{k}})) \equiv 0$ on $\mathcal{D}$, for each $j = 1,2,\ldots,q$, and $\underline{f}(u_{\underline{k}}) \to \underline{\mathring{\xi}}$ as $\underline{u}_{\underline{k}} \to \mathring{u}_{\underline{k}}$.

It is now a straightforward matter to use (2.2) to calculate the condition number of the map $M_{\underline{k},q}$ at $\mathring{u}_{\underline{k}}$. If we denote by

$$V_{\underline{k},q}(\underline{\xi}) = \begin{bmatrix} p_{k_1}(\xi_1) & p_{k_1}(\xi_2) & \cdots & p_{k_1}(\xi_q) \\ p_{k_2}(\xi_1) & p_{k_2}(\xi_2) & \cdots & p_{k_2}(\xi_q) \\ \cdots & \cdots & \cdots & \cdots \\ p_{k_p}(\xi_1) & p_{k_p}(\xi_2) & \cdots & p_{k_p}(\xi_q) \end{bmatrix} \in \mathbb{C}^{p\times q}$$

the "generalized Vandermonde matrix", and by $D(\underline{u};\underline{\xi})$ the diagonal matrix

$$D(\underline{u};\underline{\xi}) = \mathrm{diag}[p'(\underline{u};\xi_1),\ p'(\underline{u};\xi_2),\dots,\ p'(\underline{u};\xi_q)] \in \mathbb{C}^{q\times q}$$

(where the prime in $p'(\underline{u};x)$ indicates differentiation with respect to x), we find that

$$\mathrm{cond}(M_{\underline{k},q};\mathring{u}_{\underline{k}}) = \frac{\|\mathring{u}_{\underline{k}}\|}{\|\mathring{\underline{\xi}}\|}\,\|D^{-1}(\mathring{u};\mathring{\underline{\xi}})V^T_{\underline{k},q}(\mathring{\underline{\xi}})\| \ . \tag{4.2}$$

Specifically, if $\underline{k} = (1,2,\dots,n)$ and $q = n$, which is the extreme case of all roots being considered simultaneously, and all coefficients undergoing changes, and if we choose the ℓ_1-vector norm and subordinate matrix norm, we get

$$\mathrm{cond}_1(M_{\underline{k},n};\mathring{u}) = \frac{\sum_{k=1}^{n}|\mathring{u}_k|}{\sum_{j=1}^{n}|\mathring{\xi}_j|}\cdot\max_{1\le k\le n}\sum_{j=1}^{n}\left|\frac{p_k(\mathring{\xi}_j)}{p'(\mathring{\xi}_j)}\right|, \tag{4.2a}$$

where $p'(\mathring{\xi}_j) = p'(\mathring{u};\mathring{\xi}_j)$. This provides the most overall description of the condition of the algebraic equation (4.1), assuming all roots to be simple. The other extreme is $p = q = 1$, in which case we write $\underline{k} = k$, $\xi_1 = \xi$, and we find

$$\mathrm{cond}(M_{k,1};\mathring{u}) = \left|\frac{\mathring{u}_k p_k(\mathring{\xi})}{\mathring{\xi}\,p'(\mathring{\xi})}\right| , \quad k = 1,2,\dots,n\ . \tag{4.2b}$$

Each condition number in (4.2b) measures the sensitivity of the root $\mathring{\xi}$ to perturbations in one single (nonvanishing) coefficient, $\overset{\circ}{u}_k$, and provides the most detailed description of the condition of the root $\overset{\circ}{\xi}$. Note, in (4.2b), that only $\overset{\circ}{\xi}$ is assumed to be simple; some or all of the other roots may well be multiple.

A reasonable compromise between (4.2a) and (4.2b) for characterizing the condition of a single root, $\overset{\circ}{\xi}$, is cond $\overset{\circ}{\xi} = \sum_{k=1}^{n} \text{cond}(M_{k,1};\overset{\circ}{u})$, that is (we drop superscripts from now on)

$$\text{cond}\ \xi = \frac{1}{|\xi\, p'(\xi)|} \sum_{k=1}^{n} |u_k p_k(\xi)|. \tag{4.3}$$

In the following, we adopt (4.3) as the condition number of the root ξ of (4.1). (Alternatively, we could use (4.2) with $q = 1$ and $\underline{k} = (1,2,\ldots,n)$.)

4.1. Equations in power form. Here, $p_k(x) = x^{k-1}$, and (4.3) assumes the form

$$\text{cond}\ \xi = \frac{1}{|\xi\, p'(\xi)|} \sum_{k=1}^{n} |u_k \xi^{k-1}|\ . \tag{4.4}$$

The condition number (4.4) is easily seen to be invariant under scaling of the independent variable by an arbitrary complex number. Denoting the zeros of $p(x)$ by $\xi_1, \xi_2, \ldots, \xi_n$, additional insight may be provided by estimating the condition of one of these, ξ_μ, in terms of all of them. A result in this vein is the inequality (Gautschi (1973))

$$\text{cond}\ \xi_\mu \le \frac{2 \prod_{\substack{\nu=1 \\ \nu\ne\mu}}^{n} \left(1 + \left|\frac{\xi_\nu}{\xi_\mu}\right|\right) - 1}{\prod_{\substack{\nu=1 \\ \nu\ne\mu}}^{n} \left|1 - \frac{\xi_\nu}{\xi_\mu}\right|}, \tag{4.5}$$

in which equality holds precisely when all zeros ξ_ν are located on a half-ray through the origin. A similar inequality, resp. equality, holds if the zeros are pairwise symmetric with respect to the origin. Note that the bound in (4.5), like the condition number itself, is invariant with respect to scaling.

We illustrate (4.5) by several examples, beginning with the well-known example due to Wilkinson of a severely ill-conditioned equation.

Example 4.1 (Wilkinson (1963, p. 41ff)): $\xi_\nu = \nu$, $\nu=1,2,\ldots,n$.

This is a root configuration for which (4.5) holds with equality sign. There follows, by a simple computation,

$$\operatorname{cond} \xi_\mu = \frac{(\mu+n)! - \mu^n \mu!}{\mu!^2 (n-\mu)!}, \quad \mu = 1,2,\ldots,n.$$

An asymptotic analysis for large n will show that the worst conditioned root is the one near $n/\sqrt{2} = .7071\ldots n$. (For $n = 20$, the case considered by Wilkinson, the distinction goes to $\xi_{14} = 14$.) Its condition number grows exponentially,

$$\max_{1\le\mu\le n} \operatorname{cond} \xi_\mu \sim \frac{1}{\pi(2-\sqrt{2})n} \left(\frac{\sqrt{2}+1}{\sqrt{2}-1}\right)^n, \quad n \to \infty. \tag{4.6}$$

The best conditioned root is $\xi_1 = 1$, with a condition number that grows very slowly,

$$\operatorname{cond} \xi_1 \sim n^2, \quad n \to \infty. \tag{4.7}$$

It is instructive to observe what happens if one of the coefficients u_k in

$$\prod_{\nu=1}^{n} (x-\nu) = x^n + u_n x^{n-1} + \ldots + u_1,$$

say the coefficient u_{k_0}, $k_0 = [(n+3)/2]$, is continuously perturbed,

$$u_{k_0}(t) = (1+t)u_{k_0}, \quad 0 \le t \le \varepsilon,$$

all other coefficients being held constant. The resulting motion of the roots† is shown in Figure 4.1 for $n = 10$, $\varepsilon = 8\times10^{-7}$, and in Figure 4.2 for the ten "most active" roots in the case $n = 20$, $\varepsilon = 7\times10^{-14}$ (!). Initially, of course, the zeros are all confined to move along the real axis. Not before long, however, a number of them will collide, each time branching off into pairs of conjugate complex roots. When $n = 10$, there are three collisions within $0 \le t \le \varepsilon$, occuring at $t_1 = 1.02567\times10^{-7}$, $t_2 = 1.53420\times10^{-7}$, and $t_3 = 7.21568\times10^{-7}$,

†The graphs were obtained by numerical integration of the differential equations satisfied by $\xi_\nu(t)$, $\nu = 1,2,\ldots,n$. The exact instances of collision were determined by finding the t-zeros of the resultant of p and p'. The graphs after each collision represent the absolute values of the conjugate complex roots produced by the collision.

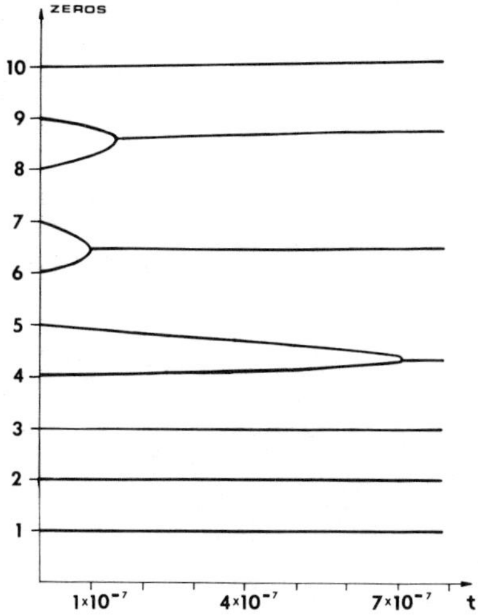

Figure 4.1
Root paths for Example 4.1
n=10

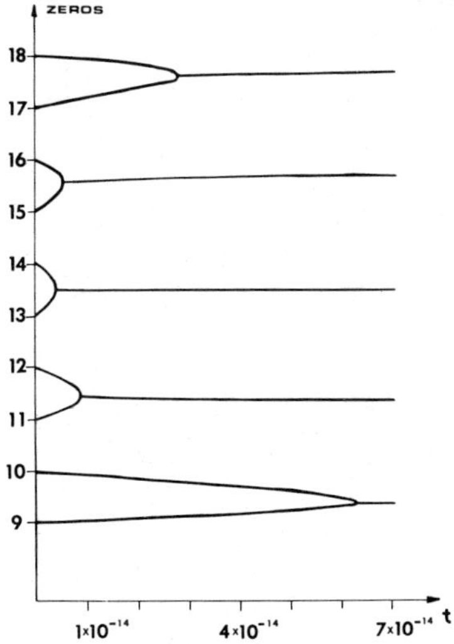

Figure 4.2
Root paths for Example 4.1
n=20

and one further collision (not shown in Figure 4.1) at $t_4 = 1.17328\times 10^{-4}$. These are the only collisions in $0 \leq t \leq 1$, as far as we could determine. For $n = 20$, there are five collisions within $0 \leq t \leq \varepsilon$, at approximately $t_1 = 4.0\times 10^{-15}$, $t_2 = 5.4\times 10^{-15}$, $t_3 = 8.9\times 10^{-15}$, $t_4 = 2.75\times 10^{-14}$, $t_5 = 6.25\times 10^{-14}$, and several more later on (e.g., at $t_6 = 1.01\times 10^{-12}$, $t_7 = 1.67\times 10^{-12}$).

The behavior in Figures 4.1 and 4.2 may be viewed as an elementary example of a bifurcation phenomenon (catastrophe theory, if you will), the special feature, here, being the almost infinitesimal time scale on which the phenomenon takes place.

Example 4.2 (Wilkinson (1963, p. 44ff)): $\xi_\nu = 2^{-\nu}$, $\nu=1,2,\ldots,n$.

The roots ξ_ν accumulate rapidly near the origin, which at first might suggest that they become more and more ill-conditioned. In reality, however, they are all quite well-conditioned. This can be seen from the inequality

$$\operatorname{cond} \xi_\mu < 2 \prod_{\substack{\nu=1 \\ \nu\neq\mu}}^{n} \frac{1 + \left|\frac{\xi_\nu}{\xi_\mu}\right|}{\left|1 - \frac{\xi_\nu}{\xi_\mu}\right|}, \tag{4.8}$$

which follows at once from (4.5), and which in the case at hand yields

$$\text{cond } \xi_\mu < 2 \prod_{\nu=1}^{\infty} \left(\frac{1+2^{-\nu}}{1-2^{-\nu}}\right)^2 = 136.32\ldots \quad . \tag{4.9}$$

The condition is thus bounded by a relatively small number (as condition numbers go), uniformly in n.

Since the bound in (4.8) is invariant with respect to reciprocation, the same result holds for the roots $\xi_\nu = 2^\nu$, $\nu = 1,2,\ldots,n$.

Example 4.3 (Roots of unity): $\xi_\nu = e^{2\pi i\nu/n}$, $\nu = 1,2,\ldots,n$.

Since $p(x) = x^n-1$, Equation (4.4) gives at once

$$\text{cond } \xi_\mu = \frac{1}{n}, \quad \mu = 1,2,\ldots,n. \tag{4.10}$$

All roots are equally well-conditioned, the condition in fact getting better with increasing degree! The example, of course, is quite trivial, and (4.10) is just another way of saying that $(1+\varepsilon)^{1/n} - 1 \sim \varepsilon/n$ as $\varepsilon \to 0$.

4.2. Equations expressed in terms of orthogonal polynomials. We now assume that the equation is written in the form

$$p(x) = 0, \quad p(x) = \pi_n(x) + \sum_{k=1}^{n} u_k \pi_{k-1}(x), \tag{4.11}$$

where $\{\pi_r\}$ is a set of orthogonal polynomials. It is to be noted that the normalization in (4.11) is such that $p(x)$ and $\pi_n(x)$ have the same leading coefficient, if expressed in powers of x. The condition number of any (simple) root ξ of (4.11) is thus

$$\text{cond } \xi = \frac{1}{|\xi p'(\xi)|} \sum_{k=1}^{n} |u_k \pi_{k-1}(\xi)| . \tag{4.12}$$

It is easily seen that cond ξ does not depend on the particular way the orthogonal polynomials π_r are normalized. Note, however, again, that changing the normalization of π_n also changes p, according to the normalization of the equation adopted in (4.11).

An easy lower bound can be had by noting that $\sum_{k=1}^{n} |u_k \pi_{k-1}(\xi)| \geq |\sum_{k=1}^{n} u_k \pi_{k-1}(\xi)| = |\pi_n(\xi)|$. Thus,

$$\operatorname{cond} \xi \geq \left|\frac{\pi_n(\xi)}{\xi p'(\xi)}\right| . \tag{4.13}$$

For an upper bound we could apply Schwarz' inequality to the sum in (4.12), but the result is not particularly revealing. It appears difficult, indeed, to extract from (4.12) much detailed information concerning the qualitative behavior of the condition of ξ. We may note, however, that there are three factors which influence its magnitude: (i) the magnitude of the Fourier coefficients u_k of p; (ii) the magnitude of the orthogonal polynomials π_k evaluated at the root ξ; (iii) the magnitude of $\xi p'(\xi)$. Since orthogonal polynomials grow rapidly outside their interval of orthogonality, it seems imperative, in view of (i) and (ii), that the interval of orthogonality be selected such as to contain ξ, if ξ is real.

It is quite possible that equations that are ill-conditioned in power form become well-conditioned when expanded in orthogonal polynomials, and vice versa. We can see this already by reexamining the examples discussed previously (cf. Gautschi (1973)). We begin with Wilkinson's example, whose roots we now scale to be enclosed in the interval $[0,1]$. As we have noted earlier, such a scaling does not affect the condition of the roots, if the equation is in power form.

Example 4.1': $\xi_\nu = \nu/n$, $\nu = 1,2,\ldots,n$.

The condition number (4.12) can be computed for various (classical) polynomials $\{\pi_r\}$ orthogonal on $[0,1]$. It turns out that the Chebyshev polynomials of the second kind perform best (in the sense of making $\max_\mu \operatorname{cond} \xi_\mu$ smallest). Some numerical results are shown in the second column of Table 4.1. They are contrasted in the third column with the analogous condition numbers for the equation in power form (see Example 4.1). The improvement is clearly significant. We remark, nevertheless, that the condition still grows exponentially with n (though at a moderate rate), as can be deduced from the inequality (4.13); if n is even, e.g., one finds

$$\max_\mu \operatorname{cond} \xi_\mu \geq \frac{(n/4)^n}{(n/2)!^2} \sim \frac{1}{\pi n} (e/2)^n , \quad n \to \infty.$$

n	$\max_{\mu} \operatorname{cond} \xi_\mu$	
	Example 4.1'	Example 4.1
5	1.85	5.87×10^{2}
10	2.64×10^{1}	2.32×10^{6}
15	6.35×10^{2}	1.05×10^{10}
20	1.40×10^{4}	5.40×10^{13}

Table 4.1
The condition of the roots in Examples 4.1' and 4.1

Example 4.2': $\xi_\nu = 2\cdot2^{-\nu}$, $\nu = 1,2,\ldots,n$.

All orthogonal (on [0,1]) polynomials $\{\pi_r\}$ tried on this example led to condition numbers that grow extremely rapidly with n . For the "best" of these, the Chebyshev polynomials of the first kind, the results are shown in the second column of Table 4.2. The third column again contains the condition numbers of the same roots for the equation in

n	$\max_{\mu} \operatorname{cond} \xi_\mu$	
	Example 4.2'	Example 4.2
5	5.03×10^{1}	4.91×10^{1}
10	1.44×10^{12}	1.13×10^{2}
15	1.19×10^{30}	1.32×10^{2}
20	3.58×10^{55}	1.36×10^{2}

Table 4.2
The condition of the roots in Examples 4.2' and 4.2

power form. The contrast is striking!

It is not difficult to identify the culprit in Example 4.2': it is the derivative of p at ξ_ν, which becomes extremely small as ν approaches n . Indeed, if p is normalized to have leading coefficient one, $p(x) = x^n + \ldots$, one finds for $\nu = n$ that

$$|\xi_n p'(\xi_n)| \sim \frac{.28879}{2^{n(n-1)/2}}, \quad n \to \infty .$$

The Fourier coefficients u_k in (4.11), although reasonably small (of order 10^{-3}), are no match for this kind of decay! In the case of the power basis, the small denominator $|\xi_n p'(\xi_n)|$ in (4.4) is neutralized by an equally small numerator,

$$\sum_{k=1}^{n} |u_k \xi_n^{k-1}| \sim \frac{2.3842}{2^{(n+1)(n-2)/2}}, \qquad n \to \infty.$$

<u>Example 4.3':</u> $\xi_\nu = e^{2\pi i\nu/n}$, $\nu = 1,2,\ldots,n$.

The roots of unity, extremely well-conditioned in the power basis (Example 4.3), continue to be quite well-conditioned in orthogonal bases, provided the interval of orthogonality is reasonably chosen. The most natural choice is $[-1,1]$, and all (classical) polynomials orthogonal on this interval do quite well, yielding condition numbers ranging from about .5 for $n = 5$ to about 35 for $n = 20$.

Just to show how an unreasonable choice of orthogonality interval may turn even the roots of unity into poorly conditioned roots, consider the case of Laguerre polynomials (orthogonal on $(0,\infty)$). Here,

$$\frac{(-1)^n}{n!}(x^n-1) = (-1)^n\left(1-\frac{1}{n!}\right) + \sum_{r=1}^{n} (-1)^{n-r} \binom{n}{r} L_r(x), \quad (4.14)$$

as can be derived from an integral formula for Laguerre polynomials (cf. Buchholz (1953, p. 120, Eq. (4β))). Therefore, from (4.12), one gets

$$\text{cond}\ \xi_\mu = (n-1)!\left(1-\frac{1}{n!} + \sum_{r=1}^{n-1} \binom{n}{r} |L_r(\xi_\mu)|\right).$$

Some numerical values are shown in Table 4.3; they speak for themselves!

n	$\max_{\mu} \text{cond}\ \xi_\mu$
5	3.17×10^{3}
10	5.45×10^{9}
20	9.71×10^{24}
40	1.92×10^{61}

Table 4.3.

The condition of the roots of unity in the Laguerre polynomial basis

The case $n = 10$ is further illustrated in Figure 4.3, which shows the motion of the roots ξ_μ induced by a multiplication of the single coefficient

$$(-1)^{n-r_0}\binom{n}{r_0}, \quad r_0 = \lfloor \frac{n+1}{2} \rfloor ,$$

in (4.14) by $1+t$ and variation of t from 0 to 10^{-8}.

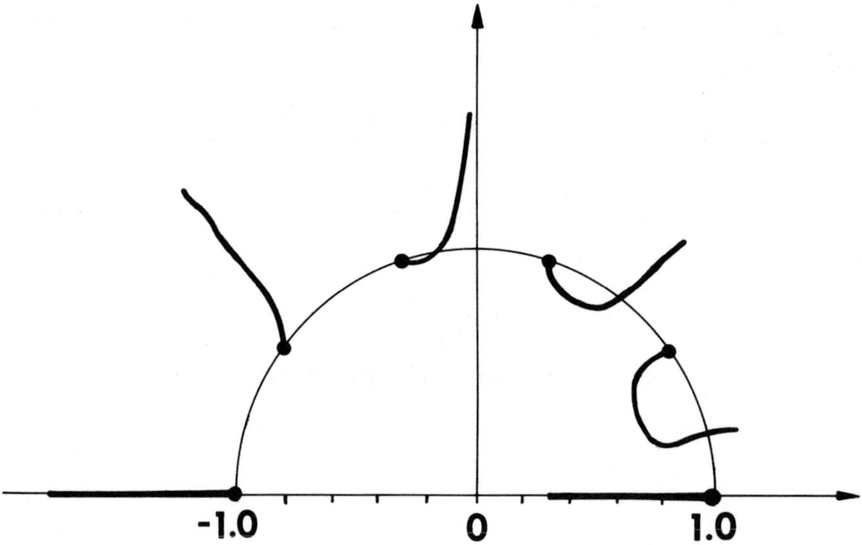

Figure 4.3

Root paths for Example 4.3'
n=10

5. GENERATION OF ORTHOGONAL POLYNOMIALS.

Generating orthogonal polynomials is fairly straightforward once the three-term recurrence relation, which we know they must satisfy, is explicitly available. Such is the case for all classical orthogonal polynomials. We are interested here in the more difficult situation where the recurrence relation is not explicitly known. Our object, indeed, is to generate it. The problem is closely related to the problem of constructing Gaussian quadrature formulas, and we use this connection to discuss the condition of the problem.

5.1. Statement of the problem. Let the desired polynomials $\{\pi_r\}$ be orthogonal on the (finite or infinite) interval (a,b) with respect to a nonnegative measure $d\sigma(x)$ on (a,b), where $\sigma(x)$ has at least $n+1$ points of increase. We assume that the first $2n$ moments of $d\sigma$ exist,

$$\mu_k = \int_a^b x^k d\sigma(x), \qquad k = 0,1,2,\ldots,2n-1. \tag{5.1}$$

It is well known that the orthogonal polynomials π_r then exist for $r = 0,1,\ldots,n,$ and satisfy a recurrence relation of the form

$$\begin{cases} \pi_{-1}(x) = 0 \ , \quad \pi_0(x) = 1 \ , \\ \pi_{k+1}(x) = (x-\alpha_k)\pi_k(x) - \beta_k\pi_{k-1}(x), \quad k=0,1,2,\ldots,n-1, \end{cases} \tag{5.2}$$

with $\beta_k > 0$. (β_0 is arbitrary, and will be set equal to 0.) Associated with (5.2) is the symmetric tridiagonal matrix of order n,

$$J_n = \begin{bmatrix} \alpha_0 & \sqrt{\beta_1} & & & 0 \\ \sqrt{\beta_1} & \alpha_1 & \sqrt{\beta_2} & & \\ & \ddots & \ddots & \ddots & \\ & & \ddots & \ddots & \sqrt{\beta_{n-1}} \\ 0 & & & \sqrt{\beta_{n-1}} & \alpha_{n-1} \end{bmatrix}, \tag{5.3}$$

the Jacobi matrix, whose r-th leading principal submatrix has the characteristic polynomial π_r, $r = 1,2,\ldots,n.$ Let $\xi_1,\xi_2,\ldots,\xi_n$ be the zeros of π_n, hence the eigenvalues of

J_n (all are simple and contained in (a,b), as is well known). Let $v_j = [v_{1j}, v_{2j}, \ldots, v_{nj}]^T$ be the normalized eigenvector of J_n belonging to the eigenvalue ξ_j,

$$J_n v_j = \xi_j v_j , \quad v_j^T v_j = 1 .$$

Then (Golub & Welsch (1969))

$$\begin{aligned} \int_a^b f(x)d\sigma(x) &= \sum_{j=1}^{n} \lambda_j f(\xi_j) + R_n(f) , \\ \lambda_j &= \mu_0 v_{1j}^2 , \quad j = 1,2,\ldots,n, \end{aligned} \tag{5.4}$$

is the Gaussian quadrature formula associated with the measure $d\sigma$, i.e., $R_n(f) = 0$ for each $f \in \mathbb{P}_{2n-1}$.

There are classical procedures for generating the Jacobi matrix J_n, hence the Gaussian quadrature formula (5.4), from the given moments μ_k in (5.1). Unfortunately, as will be seen in the next subsection, the underlying map is likely to be severely ill-conditioned. The moments μ_k, indeed, are a poor way of "codifying" the measure $d\sigma$. More promising are modified moments,

$$\nu_k = \int_a^b p_k(x)d\sigma(x), \qquad k = 0,1,2,\ldots,2n-1 , \tag{5.5}$$

where $\{p_r\}$ is a suitably selected set of polynomials (cf. Sack & Donovan (1972)). We shall assume that the p_r, like the π_r, satisfy a recurrence relation of the type

$$\begin{cases} p_{-1}(x) = 0 , \quad p_0(x) = 1 , \\ p_{k+1}(x) = (x-a_k)p_k(x) - b_k p_{k-1}(x) , \quad k=0,1,2,\ldots,n-1, \end{cases} \tag{5.6}$$

but now with coefficients a_k, b_k that are known. For example, $\{p_r\}$ may consist of a set of known (classical) orthogonal polynomials. If all a_k, b_k are zero, then $p_k(x) = x^k$, $k=0,1,2,\ldots,$ and the modified moments reduce to ordinary moments, $\nu_k = \mu_k$, $k=0,1,2,\ldots$.

The problem we wish to consider is the following: Given the modified moments ν_k in (5.5), determine the Jacobi matrix (5.3). In particular, this will also determine the orthogonal polynomials $\{\pi_r\}_{r=0}^{n}$, by virtue of (5.2), and the associated Gaussian quadrature formulas, by virtue of (5.4).

5.2. Condition of the problem. Since we need to consider perturbations in the modified moments, we temporarily denote the given measure by $d\mathring{\sigma}$, and its modified moments by $\mathring{\nu}_k$,

$$\mathring{\nu}_k = \int_a^b p_k(x)\,d\mathring{\sigma}(x) \;, \quad \mathring{\nu} = [\mathring{\nu}_0, \mathring{\nu}_1, \ldots, \mathring{\nu}_{2n-1}]^T \;.$$

The set of all $\nu = [\nu_0, \nu_1, \ldots, \nu_{2n-1}]^T$, as $d\sigma$ varies through all nonnegative measures, describes a convex cone C_n in $\mathbb{R}^{2n}$, of which $\mathring{\nu}$ is an interior point, by our assumptions on $d\sigma$. Let $\mathcal{D} \subset C_n$ be a cone generated by a small neighborhood of $\mathring{\nu}$. We consider the following two maps: The map $M_n : \mathcal{D} \subset \mathbb{R}^{2n} \to \mathbb{R}^{2n-1}$ which associates to each $\nu \in \mathcal{D}$ the vector of coefficients $[\alpha_0, \alpha_1, \ldots, \alpha_{n-1}, \beta_1, \ldots, \beta_{n-1}]^T \in \mathbb{R}^{2n-1}$ in the recurrence relation (5.2), or else, the tridiagonal elements of the Jacobi matrix J_n in (5.3), and the map $K_n : M_n \mathcal{D} \subset \mathbb{R}^{2n-1} \to \mathbb{R}^{2n}$ taking us from the Jacobi matrix (5.3) to the Gaussian nodes and weights $[\xi_1, \xi_2, \ldots, \xi_n, \lambda_1, \lambda_2, \ldots, \lambda_n]^T$. (The drop in the dimension of $M_n \mathcal{D}$ is due to the fact that M_n is homogeneous of degree zero, $M_n(t\nu) = M_n \nu$, all $t > 0$.) The map M_n corresponds to the problem of generating orthogonal polynomials from modified moments, the composite map $K_n \circ M_n$ to the problem of constructing Gaussian quadrature formulas from modified moments. The latter is equivalent to solving the system of algebraic equations

$$\sum_{j=1}^{n} \lambda_j p_k(\xi_j) = \nu_k \;, \quad k=0,1,2,\ldots,2n-1 \;. \tag{5.7}$$

We note that K_n is reasonably well-conditioned: The zeros ξ_j, as eigenvalues of the symmetric matrix J_n, are clearly well-conditioned, and so are, to a slightly lesser degree, the weights λ_j, as they are determined by the first components of the eigenvectors v_j of J_n; cf. (5.4). (The asymptotic condition numbers for the eigenvectors v_j are known to have order of magnitude $\|J_n\|_2 \sum_{i \neq j} 1/|\xi_j - \xi_i|$; see Wilkinson (1965, p. 70). Typically, this grows no faster than $O(n^2)$ for polynomials orthogonal on a finite interval; see Szegö (1975, p. 122).) We expect therefore the conditions of M_n and $K_n \circ M_n$ to be closely correlated. Certainly, by (2.3), if $K_n \circ M_n$ is ill-conditioned, then so is M_n.

While an analysis of the condition of M_n appears to be intractable, the condition of $K_n \circ M_n$, i.e., of the system of nonlinear equations (5.7), can indeed be estimated.

We begin with the classical approach via ordinary moments, $\nu = \mu$, hence $a_k = b_k = 0$ in (5.6). To be specific, we take as interval of orthogonality [0,1], and normalize μ by $\mu_0 = 1$. Writing $[\mathring{\xi}_1,\ldots,\mathring{\xi}_n,\mathring{\lambda}_1,\ldots,\mathring{\lambda}_n]^T = (K_n \circ M_n)\mathring{\mu}$, the Jacobian matrix of the map $K_n \circ M_n$ at $\mu = \mathring{\mu}$ is easily seen to be the inverse of a confluent Vandermonde matrix in the nodes $\mathring{\xi}_j$, multiplied (from the left) by the inverse of the diagonal matrix $\mathrm{diag}[1,\ldots,1,\mathring{\lambda}_1,\ldots,\mathring{\lambda}_n] \in \mathbb{R}^{2n\times 2n}$. Using two-sided estimates of the (uniform) norm of inverses of confluent Vandermonde matrices, it is possible to prove (Gautschi (1968)) that

$$\mathrm{cond}_\infty(K_n \circ M_n; \mathring{\mu}) > \frac{1}{2} \max_{1\leq j\leq n} \left[\frac{\pi_n(-1)}{\pi_n'(\mathring{\xi}_j)}\right]^2, \tag{5.8}$$

where π_n is the (desired) orthogonal polynomial of degree n with respect to the measure $d\mathring{\sigma}$. Since the point -1 is outside the interval of orthogonality [0,1], it is evident from (5.8) that the condition of $K_n \circ M_n$, hence also that of M_n, grows at least exponentially with n. An idea concerning the numerical value of the growth rate can be had by considering the (representative) example $\pi_n = T_n^*$, the "shifted" Chebyshev polynomial of the first kind, for which (5.8) yields

$$\mathrm{cond}_\infty(K_n \circ M_n; \mathring{\mu}) > \frac{(17+6\sqrt{8})^n}{64n^2} \quad (\pi_n = T_n^*). \tag{5.9}$$

The lower bound in (5.9) happens to grow at the same exponential rate as the (Turing-) condition number of the n×n-Hilbert matrix!

The situation improves significantly when [a,b] is a finite interval and we start from modified moments generated by a set of (known) polynomials orthogonal on [a,b]. The condition number $\mathrm{cond}_\infty(K_n \circ M_n; \mathring{\nu})$ can then be bounded by a complicated expression (Gautschi (1970, Theorem 2.1)), which, however, suggests polynomial, rather than exponential, growth of the condition. Thus, e.g., if $\mathring{\nu}$ is the vector of modified moments with respect to the ultraspherical polynomials $P_k^{(\beta,\beta)}$, with parameter $-\frac{1}{2} \le \beta \le 0$, and $d\sigma(x) = (1-x^2)^{\alpha}dx$ is the ultraspherical measure with $-\frac{1}{2} \le \alpha \le 0$, then

$$\mathrm{cond}_{\infty}(K_n \circ M_n;\mathring{\nu}) \le \kappa_n , \tag{5.10}$$

where

$$\kappa_n \stackrel{\bullet}{\sim} \begin{cases} n^{2(\alpha+\beta)+5} & \text{if} \quad \alpha \neq 0,\ \alpha \neq \beta \\ n^{2\beta+7} & \text{if} \quad \alpha = 0,\ \beta \neq 0 \\ n^{3\alpha+7/2} & \text{if} \quad \alpha = \beta , \end{cases} \quad \text{as} \quad n \to \infty. \tag{5.11}$$

(The symbol $\stackrel{\bullet}{\sim}$ in (5.11) means that the ratio of the two members on the left and right remains between positive bounds depending on α and β, but not on n, as $n \to \infty$.)

For infinite intervals (a,b), unfortunately, the state of affairs is not quite as favorable (Gautschi (1970, Example (iii))).

5.3. <u>An algorithm</u>. A number of algorithms have been proposed to carry out the map M_n from the modified moments ν_j to the recursion coefficients α_k, β_k (Sack & Donovan (1972), Gautschi (1970), Wheeler (1974)). We describe a particularly simple algorithm due to Wheeler (1974).

We introduce the "mixed moments"

$$\sigma_{k\ell} = \int_a^b \pi_k(x)\,p_\ell(x)\,d\sigma(x), \qquad k,\ell \ge -1 , \tag{5.12}$$

and note that, by orthogonality, $\sigma_{k\ell} = 0$ for $k > \ell$, and

$$\int_a^b \pi_k^2(x)\,d\sigma(x) = \int_a^b \pi_k(x)\,x\,p_{k-1}(x)\,d\sigma(x) = \sigma_{kk}, \quad k \ge 1 .$$

The relation $\sigma_{k+1,k-1} = 0$, therefore, together with (5.2), yields immediately $\sigma_{kk} - \beta_k\sigma_{k-1,k-1} = 0$, hence

$$\beta_k = \frac{\sigma_{kk}}{\sigma_{k-1,k-1}} , \qquad k = 1,2,3,\ldots . \tag{5.13}$$

(Recall that β_0 is arbitrary, and may be set equal to 0.) Similarly, $\sigma_{k+1,k} = 0$ gives

$$\int_a^b \pi_k(x)\ x\,p_k(x)\,d\sigma(x) - \alpha_k\sigma_{kk} - \beta_k\sigma_{k-1,k} = 0 ,$$

and using (5.6) in the form

$$xp_k(x) = p_{k+1}(x) + a_k p_k(x) + b_k p_{k-1}(x) , \tag{5.14}$$

yields $\sigma_{k,k+1} + (a_k - \alpha_k)\sigma_{kk} - \beta_k\sigma_{k-1,k} = 0$, hence, together with (5.13),

$$\begin{cases} \alpha_0 = a_0 + \dfrac{\sigma_{01}}{\sigma_{00}}, \\ \alpha_k = a_k - \dfrac{\sigma_{k-1,k}}{\sigma_{k-1,k-1}} + \dfrac{\sigma_{k,k+1}}{\sigma_{kk}}, \quad k = 1,2,3,\ldots . \end{cases} \tag{5.15}$$

The σ's in turn satisfy the recursion

$$\begin{aligned} \sigma_{k\ell} = \sigma_{k-1,\ell+1} &- (\alpha_{k-1} - a_\ell)\sigma_{k-1,\ell} \\ &- \beta_{k-1}\sigma_{k-2,\ell} + b_\ell\sigma_{k-1,\ell-1}, \end{aligned} \tag{5.16}$$

as follows from (5.2) and (5.14) (where k is replaced by ℓ). To construct orthogonal polynomials π_r of degrees $r \le n$, we thus have the following <u>algorithm of Wheeler</u> (1974)[†]:

Initialization:

$$\begin{cases} \sigma_{-1,\ell} = 0, & \ell = 1,2,\ldots,2n-2, \\ \sigma_{0,\ell} = \nu_\ell, & \ell = 0,1,\ldots,2n-1, \\ \alpha_0 = a_0 + \dfrac{\nu_1}{\nu_0}, & \\ \beta_0 = 0 . & \end{cases}$$

Continuation: For $k = 1,2,\ldots,n-1$ (5.17)

$$\begin{cases} \sigma_{k\ell} = \sigma_{k-1,\ell+1} - (\alpha_{k-1} - a_\ell)\sigma_{k-1,\ell} - \beta_{k-1}\sigma_{k-2,\ell} \\ \qquad + b_\ell\sigma_{k-1,\ell-1}, \quad \ell = k,k+1,\ldots,2n-k-1, \\ \alpha_k = a_k - \dfrac{\sigma_{k-1,k}}{\sigma_{k-1,k-1}} + \dfrac{\sigma_{k,k+1}}{\sigma_{kk}}, \\ \beta_k = \dfrac{\sigma_{kk}}{\sigma_{k-1,k-1}} . \end{cases}$$

The algorithm requires as input $\{\nu_\ell\}_{\ell=0}^{2n-1}$ and $\{a_k,b_k\}_{k=0}^{2n-2}$; it furnishes $\{\alpha_k,\beta_k\}_{k=0}^{n-1}$, hence the orthogonal polynomials $\{\pi_r\}_{r=0}^{n}$, and also, incidentally, the normalizing factors $\sigma_{kk} = \int_a^b \pi_k^2(x)\,d\sigma(x)$, $k \le n-1$. The number of multiplications

[†] Equation (3.4) in Wheeler (1974) is misprinted; a_k, b_k should read a_ℓ and b_ℓ, respectively.

and divisions required is $3n^2-n-1$, the number of additions, $4n^2-3n$, the algorithm thus involving $O(n^2)$ operations altogether.

The success of the algorithm (5.17), of course, depends on the ability to compute all required modified moments ν_ℓ accurately and reliably. Most frequently, these moments are obtained from recurrence relations, judiciously employed, as for example in the case of Chebyshev or Gegenbauer moments (Piessens & Branders (1973), Branders (1976), Luke (1977), Lewanowicz (1977)). Sometimes they can be computed directly in terms of special functions, or in integer form (Gautschi (1970, Examples (ii), (iii)), Wheeler & Blumstein (1972), Blue (1979), Gautschi (1979b)).

Example 5.1: $d\sigma(x) = x^\alpha \ell n(1/x)dx, \quad 0 < x \leq 1, \quad \alpha > -1$.

Here, the modified moments with respect to the shifted Legendre polynomials $p_k(x) = (k!^2/(2k)!)P_k(2x-1)$ can be obtained explicitly (Gautschi (1979b)). For example, if α is not an integer, then

$$\frac{(2\ell)!}{\ell!^2}\nu_\ell = \frac{1}{\alpha+1}\left\{\frac{1}{\alpha+1} + \sum_{k=1}^{\ell}\left(\frac{1}{\alpha+1+k} - \frac{1}{\alpha+1-k}\right)\right\}\prod_{k=1}^{\ell}\frac{\alpha+1-k}{\alpha+1+k}, \tag{5.18}$$

$$\ell = 0,1,2,\ldots .$$

(Similar formulas hold for integral α ; see Blue (1979) for $\alpha = 0$, Gautschi (1979b) for $\alpha > 0$.) The appropriate recursion coefficients for $\{p_r\}$ are

$$\begin{cases} a_k = \frac{1}{2}, & k = 0,1,2,\ldots, \\ b_k = \frac{1}{4(4-k^{-2})}, & k = 1,2,3,\ldots . \end{cases} \tag{5.19}$$

With the quantities in (5.18) and (5.19) as input, Wheeler's algorithm (5.17) now easily furnishes the recursion coefficients α_k, β_k , $0 \leq k \leq n-1$, for the orthogonal polynomials with respect to $d\sigma(x) = x^\alpha \ell n(1/x)dx$. For $\alpha = -\frac{1}{2}$, and $n = 100$, and single-precision computation on the CDC 6500 computer (approx. 14 decimal digit accuracy), the relative errors $\varepsilon(\alpha_k)$, $\varepsilon(\beta_k)$ observed in the coefficients α_k, β_k are shown in the left half of Table 5.1 for selected values of k . The right half displays the analogous results for the power

k	Legendre moments		power moments	
	$\varepsilon(\alpha_k)$	$\varepsilon(\beta_k)$	$\varepsilon(\alpha_k)$	$\varepsilon(\beta_k)$
1	4.3×10^{-14}	6.4×10^{-14}	3.1×10^{-15}	8.7×10^{-15}
5	1.2×10^{-13}	2.1×10^{-13}	2.1×10^{-9}	8.1×10^{-10}
10	6.3×10^{-14}	2.0×10^{-13}	9.4×10^{-4}	6.2×10^{-4}
15	8.4×10^{-14}	9.7×10^{-14}	1.9×10^{1}	1.3×10^{3}
20	1.4×10^{-13}	2.8×10^{-13}		
40	4.0×10^{-13}	8.3×10^{-13}		
60	8.3×10^{-15}	5.9×10^{-14}		
80	9.2×10^{-14}	1.4×10^{-13}		
99	2.3×10^{-13}	5.1×10^{-13}		

Table 5.1

Relative errors in the recursion coefficients α_k, β_k for Example 5.1

moments $\mu_\ell = (\alpha+1+\ell)^{-2}$, and $a_k = b_k = 0$, all k. In the first case, all coefficients are obtained essentially to machine precision, attesting not only to the extremely well-conditioned nature of the problem, but also to the stability of Wheeler's algorithm. In the second case, all accuracy is lost by the time k reaches 12, which confirms the severely ill-conditioned character of the problem of generating orthogonal polynomials from moments.

References

Blue, J. L. (1979): A Legendre polynomial integral, Math. Comput. 33, to appear.

de Boor, C. (1972): On calculating with B-splines, J. Approximation Theory 6, 50-62.

de Boor, C. (1976): On local linear functionals which vanish at all B-splines but one, in: Theory of Approximation with Applications (Law, A. G. & Sahney, B. N., eds.), pp. 120-145. Academic Press, New York-San Francisco-London.

de Boor, C. (1978a): personal communication.

de Boor, C. (1978b) A practical guide to splines, Springer-Verlag, to appear.

Branders, M. (1976): Application of Chebyshev polynomials in numerical integration (Flemish), Thesis, Catholic University of Leuven, Belgium.

Buchholz, H. (1953): Die konfluente hypergeometrische Funktion, Springer-Verlag, Berlin-Göttingen-Heidelberg.

Gautschi, W. (1968): Construction of Gauss-Christoffel quadrature formulas, Math. Comput. 22 , 251-270.

Gautschi, W. (1970): On the construction of Gaussian quadrature rules from modified moments, Math. Comput. 24 , 245-260.

Gautschi, W. (1972): The condition of orthogonal polynomials, Math. Comput. 26, 923-924.

Gautschi, W. (1973): On the condition of algebraic equations, Numer. Math. 21 , 405-424.

Gautschi, W. (1979a): The condition of polynomials in power form, Math. Comput. 33, to appear.

Gautschi, W. (1979b): Remark on the preceding paper "A Legendre polynomial integral" by J. L. Blue, Math. Comput. 33 , to appear.

Golub, G. H., and Welsch, J. H. (1969): Calculation of Gauss quadrature rules, Math. Comput. 23, 221-230.

Lewanowicz, S. (1977): Construction of a recurrence relation for modified moments, Rep. No. N-23, Institute of Computer Science, Wroclaw University, Wroclaw, Poland.

Luke, Y. L. (1977): Algorithms for the computation of mathematical functions, Academic Press, New York-San Francisco-London.

Lyche, T. (1978): A note on the condition numbers of the B-spline basis, J. Approximation Theory 22, 202-205.

Natanson, I. P. (1965): Constructive function theory, Vol. III, Frederick Ungar Publ. Co., New York.

Ortega, J. M., and Rheinboldt, W. C. (1970): Iterative solution of nonlinear equations in several variables, Academic Press, New York-London.

Piessens, R., and Branders, M. (1973): The evaluation and application of some modified moments, BIT 13, 443-450.

Rice, J. R. (1966): A theory of condition, SIAM J. Numer. Anal. 3 , 287-310.

Rivlin, T. J. (1974): The Chebyshev polynomials, John Wiley & Sons, London-Sydney-Toronto.

Sack, R. A., and Donovan, A. F. (1972): An algorithm for Gaussian quadrature given modified moments, Numer. Math. 18, 465-478.

Schönhage, A. (1971): Approximationstheorie, Walter de Gruyter & Co., Berlin-New York.

Stewart, G. W. (1973): Introduction to matrix computations, Academic Press, New York-London.

Szegö, G. (1975): Orthogonal polynomials, AMS Colloquium Publications, Vol. XXIII, 4th ed.

Voronovskaja, E. V. (1970): The functional method and its applications, Translations of Mathematical Monographs Vol. 28, American Mathematical Society, Providence, R.I.

Wheeler, J. C. (1974): Modified moments and Gaussian quadratures, Rocky Mountain J. Math. 4, 287-296.

Wheeler, J. C., and Blumstein, C. (1972): Modified moments for harmonic solids, Phys. Rev. B6, 4380-4382.

Wilkinson, J. H. (1963): Rounding errors in algebraic processes, Prentice-Hall, Englewood Cliffs, N.J.

Wilkinson, J. H. (1965): The algebraic eigenvalue problem, Clarendon Press, Oxford.

Supported in part by the National Science Foundation under Grant MCS 76-00842A01.

Department of Computer Sciences
Purdue University
West Lafayette, Indiana 47907

Global Homotopies and Newton Methods

Herbert B. Keller

1. INTRODUCTION.

To solve

$$f(u) = 0 \tag{1.1}$$

with $f \in C^2(\mathbb{R}^n)$ for $u \in \mathbb{R}^n$ we can consider the homotopy

$$f(u) - e^{-\alpha t} f(u^0) = 0 \ . \tag{1.2}$$

Here $\alpha > 0$ and $0 \leq t < \infty$ so that as $t \to \infty$ the solution, $u = u(t)$, if it exists for all $t \geq 0$, must approach a root of (1.1). Differentiating in (1.2) yields

$$\text{a)} \quad f'(u) \frac{du}{dt} + \alpha f(u) = 0 \ . \tag{1.3}$$

The solution of this system of nonlinear differential equations, subject to the initial conditions

$$\text{b)} \quad u(0) = u^0 \tag{1.3}$$

yields the homotopy path $u(t)$ from u^0 to a solution

$$\lim_{t \to \infty} u(t) = u^* \ . \tag{1.4}$$

Indeed if we use Euler's method on (1.3) (a poor choice from the point of view of efficiency) to approximate this path we get the sequence $\{u^\nu\}$ defined by:

ISBN 0-12-208360-1

$$f'(u^\nu)(u^{\nu+1}-u^\nu) + \Delta t_\nu \alpha f(u^\nu) = 0 . \tag{1.5}$$

With uniform net spacing $\Delta t_\nu \equiv \Delta t$ and with $\alpha \equiv (\Delta t)^{-1}$ this is precisely Newton's method to approximate a root of (1.1) starting with the initial guess u^0. Of course the indicated path does not always exist. It does exist for instance if u^* is an isolated root, i.e. $f'(u^*)$ non-singular, and $\|u^0 - u^*\|$ is sufficiently small. But if $f'(u)$ may become singular along the path defined by (1.3) the method need not converge. This is one of the basic difficulties to be circumvented in devising global Newton or rather Newton-like methods. Not unrelated is the fact that we frequently seek non-isolated roots and it would be nice to have an efficient method for them, too.

A key to devising global methods is to give up the monotone convergence implied in (1.2) (i.e. each component of $f \to 0$ monotonically in t) and to consider more general homotopies. One way to do this is by allowing $\alpha = \alpha(u)$ in (1.3). In fact Branin [2] employs

$$\text{a)} \quad \alpha(u) = \text{sign det } f'(u) , \tag{1.6}$$

and Smale [9] considers various related choices, say of the form

$$\text{b)} \quad \text{sign } \alpha(u) = \text{sign det } f'(u) . \tag{1.6}$$

Then if $f(u)$ satisfies appropriate boundary conditions on $\partial\Omega$ (see 2.3), for some bounded open set $\Omega \subset \mathbb{R}^n$, it is shown by Smale [9] that for almost all $u^0 \in \partial\Omega$ the homotopy path defined by (1.3), and (1.6) is such that

$$\lim_{t \to t_1} u(t) = u^* , \tag{1.7}$$

where $f(u^*) = 0$ and $0 < t_1 \leq \infty$. Note that with such choices for $\alpha(u)$, the corresponding schemes need not always proceed in the

"Newton direction": $-[f'(u)]^{-1}f(u)$, but frequently go in just the <u>opposite</u> direction. The switch in direction occurs whenever the Jacobian, $\det f'(u(t))$, changes sign. It turns out that singular matrices $f'(u)$ on the path $u(t)$ cause no difficulties in the proof of Smale's result. However they do cause trouble in attempts to implement this and most other global Newton methods numerically. Roughly speaking "small steps" must be taken in the neighborhood of vanishing Jacobians. This feature is not always pointed out in descriptions of the implementations but it is easily detected. We shall show how these difficulties can be eliminated (in principal and even in practice) by using a somewhat different homotopy.

Specifically we consider, for some fixed u^0,

$$G(u, \lambda) \equiv f(u(s)) - \lambda(s)\, f(u^0) = 0\ . \tag{1.8}$$

With s as parameter the smooth homotopy path $u(s)$ must satisfy the differential equation

$$f'(u)\dot{u} - \dot{\lambda}\, f(u^0) = 0\ . \tag{1.9}$$

In addition we do not specify λ in advance but impose the condition

$$\|\dot{u}(s)\|^2 + \dot{\lambda}^2(s) = 1\ . \tag{1.10}$$

This has the effect of making s the "arclength" parameter along the path $[u(s), \lambda(s)]$ in $\mathbb{R}^{n+1}$. In many global homotopy methods arclength is employed as a parameter but on the path $u(t)$ in $\mathbb{R}^n$. The difference, as we shall see, is crucial in devising efficient schemes. This basic idea of "inflating" the problem into one of higher dimension was first used in [6] for a much more general class of problems. If $\lambda(s^*) = 0$ at some point $s = s^*$ on the path then $u(s^*) = u^*$ is a root of (1.1). Further, several roots may be obtained if $\lambda(s)$ vanishes several times on a path. If Smale's Boundary Condition 2.3 is satisfied then we shall show that for almost all $u^0 \in \partial\Omega$ the initial data $(u(0), \lambda(0)) = (u^0, 1)$ and (1.9), (1.10) define a path on which $\lambda(s)$ vanishes (or approaches zero). These formal results are presented in Sections 2 and 3.

We discuss, in Section 4, practical procedures for computing the path $[u(s), \lambda(s)]$ or preferably reparametrizations of it. In particular we introduce a pseudo-arclength continuation procedure in which the parameter is distance along a local tangent ray to the path. Using this parameter we show, in Section 5, how to accurately locate the roots of (1.1) and the "limit points", where $\det f'(u) = 0$ on the path. These latter points are of great interest in many physical applications.

Very closely related methods appear in the thesis of Abbott [1] and in recent reports of Elken [4] and Garcia and Zangwill [5]. They give many references to the literature of mathematical economics and optimization where such methods play an important role. A nice survey of the very closely related work on piecewise linear homotopies is given by Eaves [3].

2. ROOTS ON THE REGULAR HOMOTOPY PATHS.

For a fixed $u^0 \in \partial\Omega$ the critical points of $G(u, \lambda)$ are defined as those in the set $C \subset \mathbb{R}^{n+1}$:

$$C \equiv \left\{(u, \lambda):\; u \in \mathbb{R}^n,\; \lambda \in \mathbb{R},\; \operatorname{rank} \frac{\partial G(u, \lambda)}{\partial (u, \lambda)} < n\right\} \tag{2.1}$$

The critical values are those in the set $G(C) \subset \mathbb{R}^n$. With the smoothness assumed on $f(u)$ we have $G(u, \lambda) \in C^2(\mathbb{R}^{n+1})$ and Sard's Theorem asserts that $\operatorname{meas} G(C) = 0$. If $z \in$ Range G and $z \notin G(C)$ then z is called a regular value of G. In this case we have:

Lemma 2.2. If z is a regular value of G then $G^{-1}(z)$ is a C^1 manifold of dimension 1 (i.e. it consists of smooth arcs or closed loops diffeomorphic to a circle).

These definitions and results are standard in differential topology, see

Milnor, [8].

We will assume in all that follows that the bounded open set $\Omega \subset \mathbb{R}^n$ has a smooth connected boundary $\partial\Omega$. Further $f(u)$ is assumed to satisfy on this boundary the

Boundary Condition 2.3 (Smale):

a) $\frac{\partial f(u)}{\partial u}$ is nonsingular on $\partial\Omega$; (2.3)

and either:

b) $\left(\frac{\partial f(u)}{\partial u}\right)^{-1} f(u)$ points into $\Omega \; \forall u \in \partial\Omega$; (2.3)

or:

c) $\left(\frac{\partial f(u)}{\partial u}\right)^{-1} f(u)$ points out of $\Omega \; \forall u \in \partial\Omega$. (2.3)

A simple argument reveals that on the surface of any sufficiently small sphere centered about an isolated root of (1.1), the boundary conditions are satisified. However we can also deduce that $f(u)$ has a zero in any Ω on which the boundary conditions are satisified. More precisely this root lies on an integral curve of (1.9), (1.10) which is the main result in

Theorem 2.4. Let $f: \overline{\Omega} \to \mathbb{R}^n$ be C^2 and satisfy the boundary condition (2.3). Then for any $u^0 \in \partial\Omega$ for which 0 is a regular value of $G(u, \lambda)$ there is a C^1 solution $[u(s), \lambda(s)]$ of (1.9), (1.10), over $0 \leq s \leq s_F$ starting at

a) $(u(0), \lambda(0)) = (u^0, 1)$, (2.4)

and terminating at $(u(s_F), \lambda(s_F))$ where:

b) $u(s_F) \in \partial\Omega, \; |\lambda(s_F)| < L \equiv \max_{x \in \overline{\Omega}} \|f(x)\| / \min_{y \in \partial\Omega} \|f(y)\|.$ (2.4)

For an odd number of points $s_\nu \in (0, s_F)$:

c) $\lambda(s_\nu) = 0$ and $f(u(s_\nu)) = 0$. (2.4)

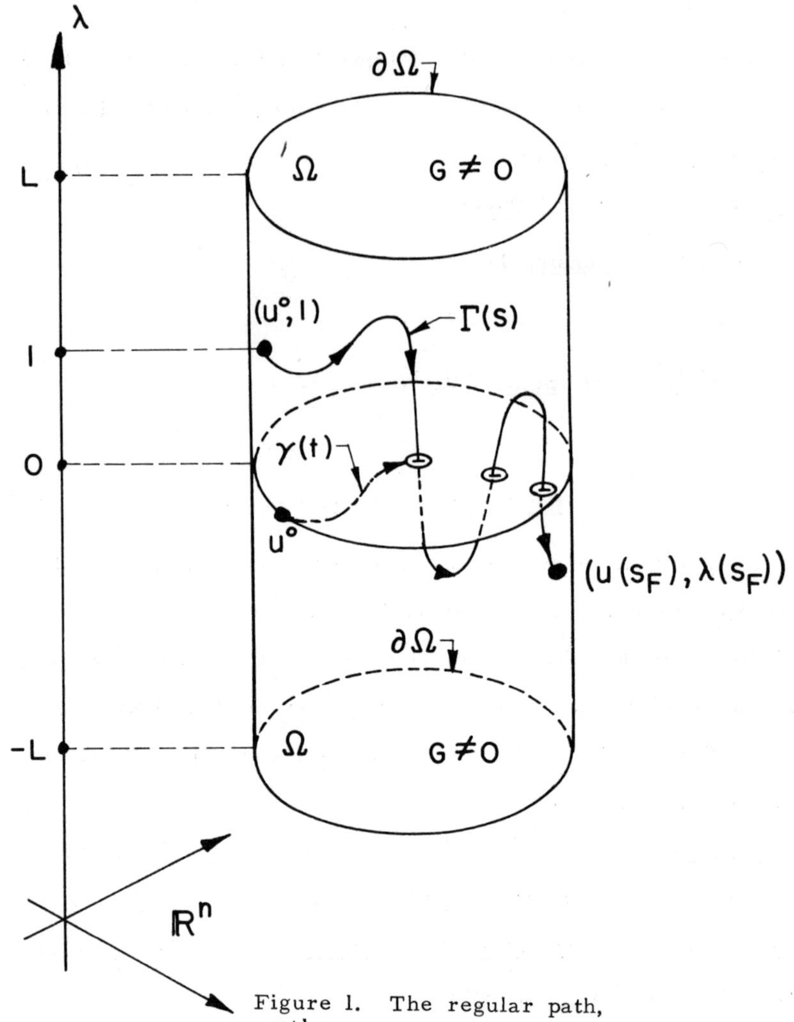

Figure 1. The regular path, $\Gamma(s) \subset \mathbb{R}^{n+1}$ and Smale's path, $\gamma(t) \subset \mathbb{R}^n$

Proof: In $\mathbb{R}^{n+1}$ we consider the cylinder

$$K \equiv \overline{\Omega} \times [-L, L] \,, \tag{2.5}$$

where L is defined in (2.4b); see Figure 1. Then for any fixed $u^0 \in \partial\Omega$ we have $G(u, \lambda) \neq 0$ on the bases of K: $\lambda = \pm L$, $u \in \overline{\Omega}$. But on the cylindrical surface of K there is at least one zero of (1.8), namely at $(u, \lambda) = (u^0, 1)$. Further, zero is assumed to be a regular value of G and clearly

$$\left.\frac{\partial G(u,\lambda)}{\partial(u,\lambda)}\right|_{(u^0,1)} = \left[\frac{\partial f(u^0)}{\partial u}, \ -f(u^0)\right] \tag{2.6}$$

where by (2.3), $\partial f(u^0)/\partial u$ is nonsingular. Thus Lemma 2.2 implies that there is a C^1 component of $G^{-1}(0)$ containing the point $(u^0, 1)$. We denote this component by $\Gamma(s) \equiv [u(s), \lambda(s)]$ and it clearly satisfies (1.8), (1.9) and (2.4a). Adjoining (1.10) simply makes s arclength on $\Gamma(s)$. The sign of the tangent vector $(\dot{u}(0), \dot{\lambda}(0))$ to $\Gamma(s)$ at $s = 0$ is chosen to make $s > 0$ for $\Gamma(s) \in K$. That is, $\dot{u}(0)$ at u^0 points <u>into</u> Ω. Then continuity along the component $\Gamma(s)$ determines the unique orientation of the tangent $(\dot{u}(s), \dot{\lambda}(s))$ satisfying (1.9) and (1.10).

The path $\Gamma(s)$ for $s > 0$ cannot meet the bases of K since L in (2.4b) is so large that G does not vanish for $|\lambda| = L$. The path $\Gamma(s)$ cannot terminate in the interior of K since if it had an interior limit point that point must lie on Γ. Then by the implicit function theorem the path could be continued (with a positive orientation) since it consists of regular points. Thus $\Gamma(s)$ must meet the cylindrical surface of K for some $s = s_F > 0$. Since the tangent $(\dot{u}(s_F), \dot{\lambda}(s_F))$ to $\Gamma(s)$ at s_F cannot point into K it follows that $\dot{u}(s_F)$ cannot point into Ω at $u(s_F) \in \partial\Omega$.

Recalling that both (1.8) and (1.9) hold on $\Gamma(s)$ it follows that

$$\dot{u}(s) = \frac{\dot{\lambda}(s)}{\lambda(s)} (f'(u(s))^{-1} f(u(s)), \quad \text{if} \quad \lambda(s)\det f'(u(s)) \neq 0 \ . \tag{2.7}$$

Applying this relation for $s = 0$ and $s = s_F$, we deduce, with the aid of (2.3) that

$$\frac{\dot{\lambda}(0)}{\lambda(0)} \cdot \frac{\dot{\lambda}(s_F)}{\lambda(s_F)} < 0 \ . \tag{2.8}$$

That is <u>both</u> $\lambda(0)\dot{u}(0)/\dot{\lambda}(0)$ and $\lambda(s_F)\dot{u}(s_F)/\dot{\lambda}(s_F)$ point out of (or into Ω. But $\dot{u}(0)$ points into Ω and $\dot{u}(s_F)$ does not, so (2.8) follows.

Now we show in Lemma 2.10 below that for $\sigma \equiv 1$ or $\sigma \equiv -1$:

$$\text{sign } \dot{\lambda}(s) = \sigma \text{ sign det } f'(u(s)) \quad \text{on} \quad \Gamma(s) . \tag{2.9}$$

Then by (2.3a) and the connectedness of $\partial\Omega$ it follows that $\dot{\lambda}(0)$ and $\dot{\lambda}(s_F)$ have the same sign. So (2.8) implies $\lambda(0)\lambda(s_F) < 0$. Thus $\lambda(s)$ must have an odd number of zeros in $0 < s < s_F$, counting multiplicities, and our theorem follows.

Note that in our proof (2.9) is, as we show below, a deduction from our homotopy and the hypothesis of Theorem 2.4. The corresponding result (1.6) must be imposed in the methods of Smale [9] and Branin [2]. We now supply the missing lemma.

Lemma 2.10. Let $z \in \mathbb{R}^n$ be a regular value for some C^2 function $P(x): \mathbb{R}^{n+1} \to \mathbb{R}^n$. Let $x(s)$, $0 \leq s \leq s_F$ be a C^1 component of $P^{-1}(z)$. Then

$$\text{a) sign } \dot{x}_j(s) \equiv \text{sign det } P^j(x(s)), \quad \forall\, s \in [0, s_F] ; \tag{2.10}$$

or else

$$\text{b) sign } x_j(s) \equiv -\text{sign det } P^j(x(s)), \quad \forall\, s \in [0, s_F] . \tag{2.10}$$

Here $P^j(x)$ is the $n \times n$ matrix obtained by deleting the jth column from $\partial P(x)/\partial x$.

Proof. Since $P(x(s)) = z$ on $0 \leq s \leq s_F$ it follows that

$$\text{a) } \left(\frac{\partial P(x(s))}{\partial x}\right) \dot{x}(s) = 0 , \quad \dot{x}(s) \neq 0 . \tag{2.11}$$

Also

$$\text{b) rank } \left(\frac{\partial P(x(s))}{\partial x}\right) = n \tag{2.11}$$

since z is assumed to be a regular value for P. We write

$$\frac{\partial P}{\partial x} \equiv (p_1, p_2, \ldots, p_{n+1}) \quad , \quad p_j \equiv \frac{\partial}{\partial x_j} P$$

and then

$$P^j \equiv (p_1, \ldots p_{j-1}, p_{j+1}, \ldots, p_{n+1}) .$$

Suppose $\dot{x}_j(s) \neq 0$ for some $s \in [0, s_F]$. Then we claim that $\det P^j(s) \neq 0$. If not the n column in $P^j(s)$ are linearly dependent. But also $p_j(s)$ is a linear combination of the n columns in $P^j(s)$, by (2.11a) and $\dot{x}_j(s) \neq 0$, or else $p_j(s) \equiv 0$. In either case this would contradict (2.11b) so $\det P^j(s) \neq 0$. By continuity of $\dot{x}_j(s)$ and $P^j(s)$ it follows that

$$\dot{x}_j(s) \det P^j(s) > 0 \quad (\text{or} < 0) \tag{2.12}$$

between consecutive zeros of $\dot{x}_j(s)$. Further if $\dot{x}_j(s) = 0$ then $\det P^j(s) = 0$ since (2.11a) must have a nontrivial solution at each $s \in [0, s_F]$. We now need only show that for any fixed j the same sign in (2.12) holds on every interval over which $\dot{x}_j(s) \neq 0$.

By the above argument and (2.11b) it follows that for each $s_0 \in [0, s_F]$ there is some index $k = k(s_0)$ for which

$$\dot{x}_k(s_0) \det P^k(s_0) \neq 0.$$

This must persist for some open or half-open interval:

$$I(s_0) \equiv (s_0 - \delta(s_0),\ s_0 + \delta(s_0)) \cap [0, s_F].$$

Over this interval we can solve (2.11a) for the $\dot{x}_i(s)$, $i \neq k$, in terms of $\dot{x}_k(s)p_k(s)$. We get, using Cramer's rule:

$$\dot{x}_i(s) = (-1)^{i-k} \frac{\det P^i(s)}{\det P^k(s)} \dot{x}_k(s), \quad \text{all } i \neq k,\ s \in I(s_0). \tag{2.13}$$

Now the entire interval $[0, s_F]$ can be covered by a finite number of intervals of the form $I(s_0)$, for various s_0. Otherwise $\dot{x}(s) = 0$ for some s, contradicting (2.11a). For some index j say, let $\dot{x}_j(s)$ be nonzero on two disjoint intervals. By using (2.13) and a chain of intervals joining the two intervals in question it follows that $\dot{x}_j(s) \det P^j(s)$ has the same sign on every interval over which it does not vanish.

Obviously to apply Lemma 2.10 we use the notation

$$x \equiv (u, \lambda) \in \mathbb{R}^{n+1}, \quad P(x) \equiv G(u, \lambda) \in \mathbb{R}^n . \tag{2.14}$$

In Theorem 2.4 it is assumed that for some $u^0 \in \partial\Omega$, zero is a regular value of G. However for a large class of functions $f(u)$ it can be shown that zero is a regular value of G for almost all $u^0 \in \partial\Omega$. This result was proven by P. Percell [11] shortly after hearing the lecture on which this paper is based. A similar idea had been suggested by S. Smale (private communication). We state this result as follows:

Lemma 2.15 (Percell). *Let $f:\Omega \to \mathbb{R}^n$ be C^2 and satisfy the boundary condition (2.3). In addition let:*

$$\text{a)} \quad \text{rank} \frac{\partial f(u)}{\partial u} \geq n-1 \quad \forall u \in \Omega \ \text{ such that } f(u) = 0; \tag{2.15}$$
$$\text{b)} \quad f(u) = 0 \ \text{ for at most countably many } u \in \Omega .$$

Then for almost all $u^0 \in \partial\Omega$ zero is a regular value of $G(u, \lambda)$ defined in (1.8).

Proof Outline: The proof proceeds by first showing that for $\lambda \neq 0$, zero is a regular value of $G(u, v, \lambda) \equiv f(u)-\lambda f(v)$ on $\overline{\Omega}\times\partial\Omega\times(\mathbb{R}-\{0\})\to\mathbb{R}^n$. Then a study of the projection of the inverse image of zero onto $\partial\Omega$ yields the result. The conditions (2.15a, b) are employed only for the case $\lambda=0$, in a much simpler argument. Details of the proof are contained in [11].

3. RELATION TO SMALE'S PATHS

We note that Theorem 2.4 implies with the aide of Lemma 2.15 that the homotopy path defined in (1.8) leads to a root of (1.1) for almost all $u^0 \in \partial\Omega$. The same can be assured without the extra assumptions in Lemma 2.15. Indeed we shall show that the C^1 homotopy path defined by (1.9), (1.10) and entering the cylinder K at $(u^0, 1)$ does in fact approach a zero of $f(x)$ for almost all $u^0 \in \partial\Omega$. This is essentially Smale's result and indeed our proof is to show that his path, $x(t)$ defined below, and our $u(s)$ define the <u>same</u> curve in $\mathbb{R}^n$ from u^0 to the first zero of $f(u(s))$. Of course when Theorem 2.4 is applicable we know that the root is actually an interior point of the path and that other roots may lie on it. This facilitates the actual computation of the root as we show in Sections 4 and 5. However to carry out the computations in an efficient manner still another change of variable is introduced in Section 4.

With $f(x)$ as before Smale defines

$$g(x) \equiv \frac{f(x)}{\|f(x)\|} \tag{3.1}$$

which is C^2 on $\Omega - E \to S^{n-1}$. Here E is the set of zeros of f in Ω and S^{n-1} is the $n-1$ dimensional unit sphere. Then by Sard's theorem it follows that for almost all $u^0 \varepsilon \partial\Omega$ the value $g(u^0)$ is a regular value for g. Hence Lemma 2.2 applied to $g(x)$ implies that $g^{-1}(g(u^0))$ has a C^1 component, say $\gamma \equiv \{x(t)\}$, which by the boundary conditions (2.3), enters $\overline{\Omega}$ at u^0 and terminates on E. Since on γ we have $g(x(t)) = g(u^0)$ the path must satisfy

$$g'(t))\frac{dx}{dt} = 0\ , \quad x(0) = u^0. \tag{3.2}$$

However using (3.1) a calculation reveals that (3.2) can also be written in the equivalent form:

$$f'(x(t))\frac{dx}{dt} = \alpha(x(t)) \quad f(x(t))\ , \quad x(0) = u^0\ , \tag{3.3}$$

where $\alpha(x(t)) \in \mathbb{R}$. To use (3.3) to determine the entire path γ the arbitrary scalar function $\alpha(x)$ is <u>required</u> to satisfy:

$$\text{sign}\ \alpha(x) = \sigma\ \text{sign det}\ f'(x)\ , \tag{3.4}$$

where $\sigma \equiv 1$ if (2.3b) holds and $\sigma \equiv -1$ if (2.3c) holds. Then we have in summary of the above, [9]:

Theorem 3.5 (Smale). Let $f: \overline{\Omega} \to \mathbb{R}^n$ be C^2 and satisfy the boundary condition (2.3). Then for almost all $u^0 \in \partial\Omega$ there is a solution $x(t)$ of (3.3), (3.4) on $[0, t_1]$ and $x(t) \to E$ as $t \uparrow t_1$.

We claim that with an appropriate change of variables $x(t)$ from Theorem 3.5 and $u(s)$ from Theorem 2.4 satisfy:

$$x(t) = u(s(t)) \ , \quad 0 \le t < t_1 \ . \tag{3.6}$$

Indeed it is clear that (3.3) can be written in the equivalent form:

$$f'(x(t)) \frac{dx}{dt} = \frac{\alpha(x(t)) \|f(x(t))\|}{\|f(u^0)\|} \ f(u^0), \quad x(0) = u^0 \ . \tag{3.7}$$

Then with the variable change $s = s(t)$ defined by

$$\dot{\lambda}(s) \frac{ds}{dt} = \frac{\alpha(x(t))\|f(x(t))\|}{\|f(u^0)\|} \ , \quad s(0) = 0 \ ; \tag{3.8}$$

the result follows from (1.9). Of course it is important to note from (3.4) and (2.9) that $\dot{\lambda}(s(t))$ and $\alpha(x(t))$ vanish at the same t values and change sign together. Then $s(t)$ is monotone and so the variable change is single valued.

The stronger result from Theorem 2.4 thus implies that Smale's path can be continued through the root and possibly to other roots. Indeed the first use of (3.3), (3.4) which was by Branin [2], stressed that several roots could be obtained by using only one path.

4. COMPUTING PATHS.

An obvious procedure to locate and accurately approximate roots of (1.1) is to compute the path $[u(s), \lambda(s)]$ defined by (1.9), (1.10) and satisfying (1.8). One difficulty with such procedures is that the system of ordinary differential equations is not in standard form. However if, as we shall assume, u^0 is such that zero is a regular value of G (so the hypothesis of Theorem 2.4 holds) we can uniquely solve (1.9), (1.10) for $[\dot{u}(s), \dot{\lambda}(s)]$. This simply follows from the implicit function theorem and the following:

Lemma 4.1. Let $[u(s), \lambda(s)]$ be a regular path with tangent $[\dot{u}(s), \dot{\lambda}(s)]$ defined by (1.9), (1.10). Then the $(n+1)x(n+1)$ Jacobian matrix

$$\mathscr{A}(s) \equiv \begin{pmatrix} f'(u(s)) & -f(u^0) \\ 2\dot{u}^T(s) & 2\dot{\lambda}(s) \end{pmatrix} \tag{4.1}$$

is nonsingular on the path.

Proof: Suppose $A(s) \equiv f'(u(s))$ is nonsingular. Then writing

$$\mathscr{A}(s) = \begin{pmatrix} A(s) & 0 \\ 2\dot{u}^T(s) & 1 \end{pmatrix} \begin{pmatrix} I & -A^{-1}(s)f(u^0) \\ 0 & 2d(s) \end{pmatrix}$$

where $d(s) \equiv \dot{\lambda}(s) + \dot{u}^T(s)\, A^{-1}(s)\, f(u^0)$ we see that $\mathscr{A}(s)$ is non-singular iff $d(s) \neq 0$. But clearly $\dot{\lambda}(s) \neq 0$, or else $\dot{u}(s) = 0$ from (1.9) and then (1.10) could not hold. So using $\dot{u}(s)$ from (1.9) and (1.10) we get that

$$d(s) = \dot{\lambda}(s) + \frac{1}{\dot{\lambda}(s)}\, \dot{u}^T(s)\, \dot{u}(s) = \frac{1}{\dot{\lambda}(s)} \neq 0\,.$$

Next suppose $A(s)$ is singular. It must have rank $n-1$ and $f(u^0)$ is not in its range. If $f(u^0)$ were in the range of $A(s)$ then $\frac{\partial G}{\partial(u,\lambda)} = [A(s), -f(u^0)]$ would have rank $\leq n-1$ and so $[u(s), \lambda(s)]$ could not be a regular path. So at any such point, $\dot{\lambda}(s) = 0$ and $A(s)\dot{u}(s) = 0$ by (1.9). In fact any solution of $\mathscr{A}(s)\binom{R}{\zeta} = 0$ must have $\zeta = 0$ and $R = a\,\dot{u}(s)$ since $\dot{u}(s)$ spans the null space of $A(s)$. The last equation of the homogeneous system becomes: $\dot{u}^T(s)\, a\, \dot{u}(s) = 0$, and hence $a = 0$. That is only the trivial solution exists and so $\mathscr{A}(s)$ is nonsingular.

Of course when $\mathscr{A}(s)$ is nonsingular for some $s = s_0$, say, then by the implicit function theorem we can solve (1.9), (1.10) for unique continuous functions $(R(u), \zeta(u))$ such that $R(u(s_0)) = \dot{u}(s_0)$

and $\zeta(u(s_0)) = \dot{\lambda}(s_0)$. Further the solution of

$$\dot{u} = R(u) \quad , \quad \dot{\lambda} = \zeta(u) \, , \tag{4.2}$$

going through $[u(s_0), \lambda(s_0)]$ defines for $|s-s_0| < \delta$ an arc of solutions of (1.9), (1.10) which can be continued until $\mathscr{A}(s)$ in (4.1) becomes singular. Lemma 4.1 insures us that this will not occur on a regular path. So the indicated procedure can be used to construct the entire regular path. To start the integration we use, form (1.9), (1.10) at $s=0$:

$$\begin{aligned} &\text{a)} \ \dot{u}(0) = \dot{\lambda}(0)(f'(u^0))^{-1} f(u^0) \\ &\text{b)} \ \dot{\lambda}(0) = \pm \left(1 + \|(f'(u^0))^{-1} f(u^0)\|^2\right)^{-\frac{1}{2}} . \end{aligned} \tag{4.3}$$

The sign in (4.3b) is to be chosen such that $\dot{u}(0)$ in (4.3a) is directed <u>into</u> Ω from $u^0 \in \partial\Omega$. The boundary condition (2.3) assures us that this can be done.

Standard numerical procedures can be used, in principle, to approximate the solution of (4.2) and hence the path $[u(s), \lambda(s)]$ to any desired accuracy. Of course such guaranteed accurate schemes yield corresponding guaranteed accurate algorithms for solving (1.1). The only trouble with such schemes is that they are frequently impractical. The source of the trouble is twofold: i) truncation error growth if the path length is long, ii) roundoff which is always present and sometimes disastrous. The first difficulty can be overcome by taking small steps but this then enhances the second difficulty. A simple way to overcome these problems is to use (1.8) periodically to return to the regular path and thus not let the numerical integration of (1.9), (1.10) cause us to drift too far off the path. Notice that the singularity of $A(s) \equiv f'(u(s))$ does not play any role in the above considerations. Let us hasten to point out that numerical integration of (4.2) can be quite effective for large classes of problems. The alternative procedure we present below is just more effective for a broader class of problems.

Our alternative scheme, which we termed pseudo-arclength continuation in [6], employed (1.8), (1.9), (1.10) and an approximation to (1.10). Specifically we assume known a point $[u(s_0), \lambda(s_0)] \equiv [u_0, \lambda_0]$ on the path $\Gamma(s)$, and the corresponding tangent vector $[\dot{u}(s_0), \dot{\lambda}(s_0)] \equiv [\dot{u}_0, \dot{\lambda}_0]$. These quantities must satisfy (1.8)-(1.10). Then we seek another point $[v, \xi]$ on $\Gamma(s)$ by solving:

$$\begin{aligned} &\text{a)}\quad G(v,\xi) \equiv f(v) - \xi\, f(u^0) = 0 \\ &\text{b)}\quad N(v,\xi;\sigma) \equiv \dot{u}_0^{\,T}[v-u_0] + \dot{\lambda}_0[\xi-\lambda_0] - (\sigma - s_0) = 0 \end{aligned} \qquad (4.4)$$

Note that $[v, \xi] \neq [u_0, \lambda_0]$ provided $\sigma \neq s_0$. With σ as parameter we claim that: (4.4) <u>defines an arc</u> $[v(\sigma), \xi(\sigma)]$ <u>of</u> $\Gamma(s)$ <u>provided</u> $|\sigma - s_0|$ <u>is sufficiently small.</u> Indeed with $[v(s_0), \xi(s_0)] = [u_0, \lambda_0]$ we have a root at $\sigma = s_0$. The Jacobian of the system (4.4) evaluated at this root is just $\mathscr{A}(s_0)$ (to within a factor of 2 in the last row) and so is nonsingular. Then the implicit function theorem yields the existence of the stated arc. Obviously the parametrization of this arc is different from that of $\Gamma(s)$; it uses as parameter distance on a tangent ray rather than arclength. Our scheme is to generate a sequence of arcs as above by sequentially changing the point $[u_0, \lambda_0]$ and corresponding tangent $[\dot{u}_0, \dot{\lambda}_0]$.

A geometric interpretation of (4.4) is sketched in Figure 2. Specifically (4.4b) is just the condition that $[v, \xi]$ lie in the plane, $\Pi(\sigma)$, normal to the tangent vector to $\Gamma(s)$ at s_0 and at a distance $|\sigma - s_0|$ from the point of tangency. Condition (4.4a) requires $[v, \xi]$ to be the point of intersection of $\Gamma(s)$ with $\Pi(\sigma)$.

To compute $[v, \xi]$ we use Newton's method:

$$\begin{aligned} &\text{a)}\quad f'(v^\nu)\,\delta v^\nu - f(u^0)\,\delta\xi^\nu = -G(v^\nu, \xi^\nu) \\ &\qquad \dot{u}_0^{\,T}\,\delta v^\nu + \dot{\lambda}_0\,\delta\xi^\nu = -N(v^\nu, \xi^\nu;\sigma) \end{aligned} \qquad (4.5)$$

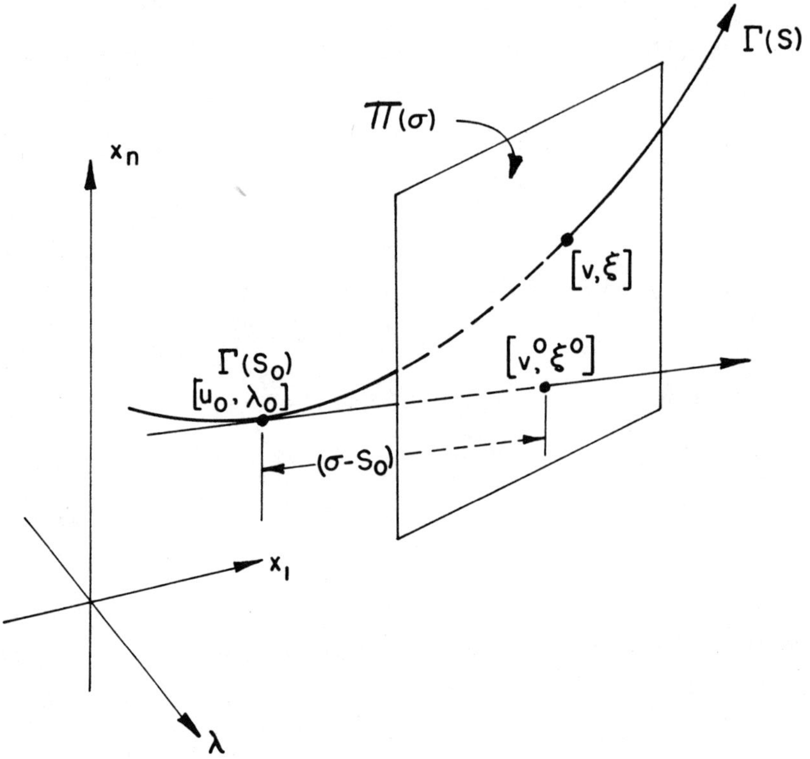

Figure 2.
Pseudoarclength continuation uses distance, $\sigma - s_0$, along a tangent ray.

The initial iterate is chosen as:

$$\text{b)} \quad \begin{aligned} v^0 &= u_0 + (\sigma - s_0)\,\dot{u}_0 \\ \xi^0 &= \lambda_0 + (\sigma - s_0)\,\dot{\lambda}_0 \end{aligned} \tag{4.5}$$

and clearly satisfies (4.4b) (see Figure 2). Of basic interest is the largest distance $|\sigma - s_0|$ for which (4.5) will converge to a point on $\Gamma(s)$. We claim that this distance can be estimated from

$$\text{a)} \quad |\sigma - s_0|^2 \, \| f''(u_0)\dot{u}_0\dot{u}_0 \| \, \mathcal{K}(\mathcal{A}(s_0)) + \mathcal{O}(|\sigma - s_0|^3) = 1 \,. \tag{4.6}$$

Here

$$\text{b)} \quad \mathcal{K}(\mathcal{A}) \equiv \|\mathcal{A}^{-1}\| \cdot \|\mathcal{A}\| \tag{4.6}$$

is the condition number of the Jacobian matrix. (It is of interest to compare (4.6) with Theorem 4.4 in [6].) The estimate (4.6) is easily obtained by writing down the Kantorovich sufficient conditions for the convergence of Newton's method (4.5a) in a sphere about $[v^0, \xi^0]$ of (4.5b). In the expansions we use the assumed continuity of $f''(u)$ in some sphere of radius $\mathcal{O}(|\sigma - s_0|)$ about u_0.

To compute the Newton iterate defined by (4.5a) we need only solve the two linear systems

$$\begin{aligned} &\text{a)} \quad f'(v^\nu)y = f(u^0)\,, \\ &\text{b)} \quad f'(v^\nu)z = G(v^\nu, \xi^\nu)\,; \end{aligned} \tag{4.7}$$

and then form:

$$\begin{aligned} &\text{c)} \quad \delta\xi^\nu = \frac{\dot{u}_0^T z - N(v^\nu, \xi^\nu; \sigma)}{\dot{u}_0^T y + \dot{\lambda}_0}\,, \\ &\text{d)} \quad \delta v^\nu = \delta\xi^\nu y - z \end{aligned} \tag{4.7}$$

Since $f'(v)$ can have at most a one dimensional null space for v near the regular path, there is no difficulty in carrying out the above calculations. At worst, i.e. if $f'(v)$ is singular, the solutions of (4.7a, b) are each equivalent to one step of inverse iteration (see [7] for more details). The only care that must be taken is to employ some form of pivoting so that the correct rank of $f'(v)$ could be determined (i.e. n or $n-1$). Even in the singular case it can be shown that this procedure converges quadratically to the root of (4.4), see Szeto [10] pp. 50-70.

After the root $[v(\sigma), \xi(\sigma)]$ of (4.4) is obtained we use it in

$$\begin{aligned} &\text{a)} \quad f'(v)\dot{v} - f(u^0)\dot{\xi} = 0 \\ &\text{b)} \quad \|\dot{v}\|^2 + |\dot{\xi}|^2 = 1 \end{aligned} \tag{4.8}$$

to solve for the root $[\dot{v}, \dot{\xi}]$ for which

$$\text{c) } \dot{u}_0^T \dot{v} + \dot{\lambda}_0 \dot{\xi} > 0 . \qquad (4.8)$$

This is easily done by first solving

$$\text{a) } f'(v)\,\phi = f(u^0) , \qquad (4.9)$$

and then forming,

$$\text{b) } \dot{v} = \dot{\xi}\,\phi , \qquad (4.9)$$
$$\text{c) } \dot{\xi} = \pm\left[1 + \|\phi\|^2\right]^{-1/2} .$$

The sign in (4.9c) is chosen to so that (4.8c) is satisfied. Note that the last iterate used in (4.7a) implies $y \doteq \phi$ so we need not even bother to solve (4.9a). (In the singular case we see that $\dot{\xi}$ is approximately zero and $\dot{v}$ is approximately a normalized null vector of $f'(v)$.) After $[v(\sigma), \xi(\sigma)]$ and $[\dot{v}(\sigma), \dot{\xi}(\sigma)]$ are computed we use them to replace $[u_0, \lambda_0]$ and $[\dot{u}_0, \dot{\lambda}_0]$ in (4.4)-(4.5) and proceed to compute a new point on $\Gamma(s)$.

Since the origin of s is arbitrary we need never change s_0. The only parameter to vary is the "steplength", $\sigma - s_0$, and we can estimate it from (4.6). This may be a very costly calculation in problems of large magnitude so alternative steplength techniques are of great interest. However since

$$f''(u_0)\dot{u}_0\dot{u}_0 = \frac{d^2}{dt^2} f(u_0 + t\dot{u}_0)\Big|_{t=0}$$

we can use the approximation:

$$\|f''(u_0)\dot{u}_0\dot{u}_0\| = \left\|\frac{1}{h^2}\left[f(u_0 + h\dot{u}_0) - 2f(u_0) + f(u_0 - h\dot{u}_0)\right]\right\| + \mathcal{O}(h^2) .$$

This requires only two extra evaluations of $f(u)$; the linear terms in $f(u)$ do not contribute and the quadratic terms are treated exactly. There are also efficient procedures for estimating the condition number of a matrix and so it may be quite practical to use (4.6) occasionally to estimate the step-length, $\sigma - s_0$.

5. LOCATING ROOTS AND LIMIT POINTS.

As indicated above we can easily generate a sequence of points $[u_0, \lambda_0]$ on $\Gamma(s)$ and corresponding tangent vectors $[\dot{u}_0, \dot{\lambda}_0]$. When λ_0 changes sign between two consecutive such points a root of $f(u) = 0$ lies on the arc of $\Gamma(s)$ between these points. Thus if $[u_0, \lambda_0] = [v(\sigma_0), \xi(\sigma_0)]$ represents the first point and $[v(\sigma_1), \xi(\sigma_1)]$ represents the second point, then the arc $[v(\sigma), \xi(\sigma)]$ defined by (4.4) with $\sigma_0 \leq \sigma \leq \sigma_1$ contains the root. We need only find $\sigma^* \epsilon (\sigma_0, \sigma_1)$ such that $\xi(\sigma^*) = 0$ and our root of (1.1) is then $u^* = v(\sigma^*)$.

It follows from the assumed smoothness of $f(u)$ that $[v(\sigma), \xi(\sigma)] \epsilon C^2[\sigma_0, \sigma_1]$. Further as stated above we have $\xi(\sigma_0)\,\xi(\sigma_1) < 0$. Then a number of standard iterative schemes can be used to compute a sequence $\{\sigma_j\}$ converging to σ^*. In particular the secant method seems attractive as it has superlinear convergence and the next iterate, σ_2, is assured to lie in (σ_0, σ_1). Specifically we have in this scheme, with $\xi_i \equiv \xi(\sigma_i)$:

$$\text{a)}\quad \sigma_{j+1} = \sigma_j - \xi_j \frac{\sigma_j - \sigma_{j-1}}{\xi_j - \xi_{j-1}} \tag{5.1}$$

The scheme can be modified to insure convergence by coupling it with the method of false position. That is we retain along with the current iterate σ_j the most recent iterate, say σ_k, such that $\xi_j \xi_k < 0$. Then if $\sigma_{j+1} \notin [\sigma_j, \sigma_k]$ we discard it and use instead

$$\text{b)}\quad \sigma_{j+1} = \frac{\xi_j \sigma_k - \xi_k \sigma_j}{\xi_j - \xi_k} \tag{5.1}$$

Whichever value $\sigma = \sigma_{j+1}$ is used in (4.4) we then need a new initial iterate $[v^0, \xi^0]$ to use in Newton's method (4.5a). Again we could simply use $\sigma = \sigma_{j+1}$ in (4.5b) but much better choices are available. Specifically we suggest using

$$v^0(\sigma_{j+1}) = \theta v(\sigma_j) + (1-\theta) v(\sigma_k) ,$$
$$\xi^0(\sigma_{j+1}) = 0 \quad , \quad \theta \equiv \frac{\sigma_{j+1} - \sigma_k}{\sigma_j - \sigma_k} . \tag{5.2}$$

Here $k = j-1$ if σ_{j+1} is determined by (5.1a), otherwise k is as in (5.1b).

In many important applications it is of interest to locate accurately limit points, where $\det f'(u(s)) = 0$ or equivalently $\dot{\lambda}(s) = 0$, on a regular path, $\Gamma(s)$. This can be done very much as with the above location of roots of $\xi(\sigma) = 0$. We examine the sequence of points $[u_0, \lambda_0]$ on $\Gamma(s)$ and corresponding tangents $[\dot{u}_0, \dot{\lambda}_0]$ to $\Gamma(s)$ and note any sign change in $\dot{\lambda}_0$. Suppose as above that this occurs in going from $[v(\sigma_0), \xi(\sigma_0)]$ to $[v(\sigma_1), \xi(\sigma_1)]$. Then on $\sigma_0 < \sigma < \sigma_1$ the arc $[v(\sigma), \xi(\sigma)]$ satisfying (4.4) has a smooth tangent $[v'(\sigma), \xi'(\sigma)] \equiv [\frac{dv(\sigma)}{d\sigma}, \frac{d\xi(\sigma)}{d\sigma}]$ which satisfies:

$$\text{a)} \quad f'(v(\sigma))v'(\sigma) - f(u^0)\, \xi'(\sigma) = 0 ,$$
$$\text{b)} \quad \dot{u}_0^T\, v'(\sigma) + \dot{\lambda}_0\, \xi'(\sigma) = 1 . \tag{5.3}$$

This tangent vector (in general not a <u>unit</u> tangent since σ is not arclength) can also be represented in terms of $\phi(\sigma)$, the solution of

$$\text{a)} \quad f'(v(\sigma))\, \phi(\sigma) = f(u^0) ; \tag{5.4}$$

with

$$\text{b)} \quad v'(\sigma) = \xi'(\sigma)\, \phi(\sigma)$$
$$\text{c)} \quad \xi'(\sigma) = [\dot{\lambda}_0 + \dot{u}_0^T\, \phi(\sigma)]^{-1} \tag{5.4}$$

A comparison of (5.4) with (4.9) and (4.8) shows that the arclength parameter, s, and pseudoarclength parameter σ are related on the arc in question by a change of variable $\sigma = \sigma(s)$ with:

$$\frac{d\sigma}{ds} = \begin{cases} |\dot{\lambda}_0 + \dot{u}_0^T \phi(\sigma)| / \sqrt{1 + \|\phi(\sigma)\|^2} & \text{if } \det f'(v(\sigma)) \neq 0 , \\ |\dot{u}_0^T \phi(\sigma)| / \|\phi(\sigma)\| & \text{if } \det f'(v(\sigma)) = 0 . \end{cases} \tag{5.5}$$

In the singular case $\phi(\sigma)$ is any null vector of $f'(v(\sigma))$. We are thus assured that $\dot{\xi}(\sigma) = \xi'(\sigma)\, d\sigma/ds$ has the same sign as $\xi'(\sigma)$, since $d\sigma/ds > 0$. Hence we are justified in seeking a root of $\xi'(\sigma) = 0$. This can now be done in exact analogy with the search for roots of $\xi(\sigma) = 0$, so we do not spell out the details. In (5.1a, b) we need only replace the $\xi_i = \xi(\sigma_i)$ by the $\xi_i' = \xi'(\sigma_i)$ and (5.2) is unaltered. The only additional calculation required is to evaluate $\xi'(\sigma_j)$ at each iterate in the sequence $\{\sigma_j\}$. But this is essentially free using (5.4c) and the previously observed relation between (5.4a) and the last Newton iterate in (4.7a).

A final observation, based on the ease in computing $\xi'(\sigma)$, is that Newton's method could easily be applied to find the root of $\xi(\sigma) = 0$. Even better, third order methods, say using Hermite interpolation, could just as easily be used. The same is not true in seeking roots of $\xi'(\sigma) = 0$ since $\xi''(\sigma)$ is not easily available.

REFERENCES

[1] J. P. Abbott, Numerical continuation methods for nonlinear equations and bifurcation problems, Thesis, Australian National University, 1977.

[2] F. H. Branin, Jr., Widely convergent method for finding multiple solutions of simultaneous nonlinear equations, I. B. M. J. Research Develop. 16 (1972) 504-522.

[3] B. C. Eaves, A short course in solving equations with P L Homotopies, in Nonlinear Programming, SIAM-AMS Proceedings, Vol. IX (1976) 73-144.

[4] T. R. Elken, On the solution of nonlinear equations by path methods, Tech. Rep. SOL77-25, Stanford University, Systems Optimization Laboratory (1977).

[5] C. B. Garcia and W. I. Zangwill, Global continuation methods for finding all solutions to polynomial systems of equations in N variables, Report 7755, University of Chicago, Center for Math. Studies in Business and Economics (1977).

[6] H. B. Keller, Numerical solution of bifurcation and nonlinear eigenvalue problems, in Applications of Bifurcation Theory (ed. by P. Rabinowitz), Academic Press, New York (1977) 359-384.

[7] H. B. Keller, Singular systems, inverse iteration and least squares, submitted to J. Lin. Alg. and Appl. 1978.

[8] J. Milnor, Topology From the Differentiable Viewpoint, University Press of Virginia, Charlotte, VA., 1965.

[9] S. Smale, A convergent process of price adjustment and global Newton methods, J. Math. Econ. 3 (1976) 107-120.

[10] R. K.-H. Szeto, The flow between rotating coaxial disks, Thesis, California Institute of Technology, 1978.

[11] P. Percell, Note on a global homotopy, draft manuscript, Univ. Houston, Texas, 1978.

Acknowledgments

This work is supported by the Department of Energy under contract EY-76-S-03-0767 Project Agreement No. 12 and by the Army Research Office under contract DAAG 29-78-C-0011.

Applied Mathematics 101-50
California Institute of Technology
Pasadena, California 91125

Problems with Different Time Scales

Heinz-Otto Kreiss

1. Introduction.

The perhaps simplest problem with different time scales is given by the initial value problem for the ordinary differential equation

(1.1) $$\varepsilon dy/dt = ay + e^{it}, \quad y(0) = y_0, \quad t \leq 0 .$$

Here ε with $0 < \varepsilon << 1$ is a small positive constant and a is a constant with Real $a \leq 0$ and $|a| \sim 1$. The solution of (1.1) is given by

(1.2) $$y = y_1(t) + y_2(t),$$

where

$$y_1(t) = -\frac{1}{a-i\varepsilon} e^{it}, \quad y_2(t) = (y_0 + \frac{1}{a-i\varepsilon})e^{(a/\varepsilon)t},$$

denote the forced and transient solution respectively.

There are two fundamentally different situations.

1) Real $a \sim -1$. Then the transient solution decays rapidly and out side a boundary layer the solution depends essentially only on the slow scale. Many people have developed methods to solve this problem (see for example [3]) and we shall not consider this case.

2) Real $a = 0$, i.e. $a = i\beta$. In general the solution of our problem will oscillate rapidly for all times except in the case that the initial values are chosen such that

$$y_2(t) \equiv 0 , \quad \text{i.e.} \quad y_0 = -(a-i\varepsilon)^{-1} .$$

In many applications one is not interested in the fast time scale. Therefore one is interested to choose the initial data such that the fast time scale is not activated. In this paper we want to discuss a general theory which leads to a systematic way to prepare the initial data. The

ISBN 0-12-208360-1

principle is extremely simple and is based on the following observation. If the solution $y(t)$ varies slowly then $d^\nu y(t)/dt^\nu \sim O(1)$ for $\nu=0,1,2,\ldots,p-1$, where $p > 1$ is a suitable number. Our principle is

<u>Choose the initial values y_0 such that for $t = 0$</u>

(1.3) $$\underline{d^\nu y/dt^\nu \sim O(1) \quad , \quad \nu=0,1,2,\ldots,p-1.}$$

Using the differential equation we can express the derivatives at $t=0$ by $y(0)$. Therefore we can determine $y(0)$ such that (1.3) is satisfied without solving the differential equations.

Let us apply the principle to the problem (1.1). For $t=0$ we have

$$\varepsilon dy/dt = ay_0 + 1 \quad , \quad t=0.$$

Thus $|dy/dt| \sim O(1)$ if and only if

$$y_0 = -\frac{1}{a} + \varepsilon\tilde{y}_0 \quad , \quad |\tilde{y}_0| \sim O(1) \quad .$$

For the second derivative we obtain at $t=0$

$$\varepsilon d^2y/dt^2 = ady/dt + i = a\tilde{y}_0 + i \quad .$$

Thus $|d^2y/dt^2| \sim O(1)$ if and only if

(1.4) $$\tilde{y}_0 = -i/a + \varepsilon\tilde{\tilde{y}}_0 \quad , \quad \text{i.e.} \quad y_0 = -\frac{1}{a}(1+i\varepsilon+O(\varepsilon^2)) \quad .$$

(1.4) gives us the first two terms in the expension of $-(a-i\varepsilon)^{-1}$ with respect to ε. Thus $y_2(0) = y_0 + (a-i\varepsilon)^{-1} = O(\varepsilon^2)$. Therefore our principle works.

In [4] we have considered general systems of ordinary differential equations and we shall discuss the results in the next section. One of the main question is the length of the time interval $0 \le t \le T$ in which one can neglect the component due to the fast time scale. This depends very much on the smoothness of the forcing function and (in the nonlinear case) on p. Consider, for example, the differential equation

$$\varepsilon dy/dt = iy + iF \quad , \quad F = e^{it} + be^{(i/\varepsilon)t} \quad .$$

Its general solution is given by

$$y = -(1-\varepsilon)^{-1}e^{it} + ((b/\varepsilon)t-c)e^{(i/\varepsilon)t} \quad ,$$

where c is an arbitrary constant. If $b \sim 1$ then F has no derivatives bounded independently of ε and the solution oscillates rapidly and its amplitude grows like $(b/\varepsilon)t$. If $F(t)$ has p derivatives bounded independently of ε then necessarily $b = O(\varepsilon^p)$. Furthermore if we choose the intial values $y(0) = y_0$ such that the solution has $p-1$ derivatives bounded independently of ε at $t=0$ then also $c = O(\varepsilon^{p-1})$. Thus we can neglect the component due to the fast scale as long as $|\varepsilon^{p-1}t| << 1$. This behavior is typical for rather general systems of ordinary differential equations.

The principle breaks down if turning points are present. Consider, for example, the differential equation

$$\varepsilon y' = i(t-1/2)y + i\varepsilon^{1/2} \quad , \quad y(0) = y_0 \quad , \quad t \geq 0. \tag{1.5}$$

Applying our principle we can choose y_0 such that $y(t)$ is smooth for $0 \leq t \leq 1/2 - \delta$. In fact, $y \approx -\varepsilon^{1/2}/(t-1/2)$. However, after we have passed the turning point $t = 1/2$ the solution of (1.5) contains a fast component

$$w = e^{\frac{i(t-1/2)^2}{2\varepsilon}} w_0 \quad , \quad |w_0| \sim O(1).$$

There is an important difference between systems

$$dy/dt = A(t)y + F(t)$$

(and correspondingly for nonlinear systems) where the fast scale is represented by eigenvalues of A which are real and negative or which are purely imaginary. In the first case turning points are often present i.e. the eigenvalue change order of magnitude. Then there is an interior boundary layer where the solution changes rapidly. The main difficulty to solve these problems is to detect the boundary layers. Once one has passed the boundary layer the fast scale is damped out. In the second case the fast scale is not damped. Luckily this situation is rather uncommon.

In section 3 we generalize the theory to hyperbolic systems

$$\begin{aligned} \partial u/\partial t &= \frac{1}{2} \sum_{j=1}^{s} \partial(A_j u)/\partial x_j + A_j \partial u/\partial x_j + F = \\ &= P(x,t,\partial/\partial x)u + F . \end{aligned} \tag{1.6}$$

Here $u = u(x,t) = (u^{(1)},\ldots,u^{(n)})'$, $F = (F^{(1)},\ldots,F^{(n)})'$ are vector functions with **n** components and $A_j(x,t)$ are real symmetric $n \times n$ matrices which depend smoothly on x,t. If the matrices A_j are constant and we want to solve the Cauchy problem then we can apply the theory for ordinary differential equations immediately because we can Fourier transform (1.6) and obtain

$$(1.7)\qquad d\hat{u}/dt = |\omega|P(i\omega')\hat{u} + \hat{F}, \quad P(i\omega') = i\sum_{j=1}^{s} A_j\omega_j', \quad \omega_j'=\omega_j/|\omega| .$$

For every fixed ω' the system (1.7) represents a systems of ordinary differential equations. Therefore we can apply our earlier results if the eigenvalues $\kappa(\omega')$ of $P(i\omega')$ split into two groupts M_1, M_2 with

$$(1.8)\qquad \begin{aligned} &\text{If } \kappa \in M_1 \text{ then } |\kappa| \geq \varepsilon^{-1}, \\ &\text{If } \kappa \in M_2 \text{ then } |\kappa| \leq c, \end{aligned}$$

where ε , c are constants with $0 < \varepsilon << 1$, $0 < c << \varepsilon^{-1}$.

In applications the system (1.6) can be written as

$$(1.9)\qquad P(x,t,\partial/\partial x) = \varepsilon^{-1}P_0(x,t,\partial/\partial x) + P_1(x,t,\partial/\partial x),$$

where the coefficients of P_0, P_1 are of order $O(1)$ and smooth. The relations (1.8) can then be expressed as

Assumption 1.1. Let

$$P_0u = \frac{1}{2}\sum \partial(\tilde{A}_ju)\partial x + \tilde{A}_j\partial u/\partial x_j = \sum \tilde{A}_j\ \partial u/\partial x_j + Du.$$

The eigenvalues $\kappa = \kappa(x,t,\omega')$ of the symbol $i\sum \tilde{A}_j(x,t)\omega_j'$ split into two groups M_1, M_2 with:

$$\text{If } \kappa \in M_1 \text{ then } |\kappa| \geq 1, \text{ if } \kappa \in M_2 \text{ then } \kappa = 0 .$$

Thus the number of eigenvalues $\kappa \in M_1$ is independent of ω.

A typical example are the shallow water equations.

$$(1.10)\qquad \frac{\partial}{\partial t}\begin{pmatrix} h \\ u \\ v \end{pmatrix} = \sqrt{gH}\begin{pmatrix} 0 & 1 & 0 \\ 1 & 0 & 0 \\ 0 & 0 & 0 \end{pmatrix}\frac{\partial}{\partial x}\begin{pmatrix} h \\ u \\ v \end{pmatrix} + \begin{pmatrix} 0 & 0 & 1 \\ 0 & 0 & 0 \\ 1 & 0 & 0 \end{pmatrix}\frac{\partial}{\partial y}\begin{pmatrix} h \\ u \\ v \end{pmatrix} + U\frac{\partial}{\partial x}\begin{pmatrix} h \\ u \\ v \end{pmatrix} + V\frac{\partial}{\partial y}\begin{pmatrix} h \\ u \\ v \end{pmatrix},$$

where H denotes the mean depth, U, V mean velocities in the x and y

direction respectively, h the deviation from the mean depth and u, v the deviation from the mean velocities. In many applications, for example, oceanography and meteorology

$$\sqrt{gH} \gg |U| + |V| \quad .$$

Then there are two different time scales present. The operator P_0 is given by

$$P_0 = \begin{pmatrix} 0 & 1 & 0 \\ 1 & 0 & 0 \\ 0 & 0 & 0 \end{pmatrix} \frac{\partial}{\partial x} + \begin{pmatrix} 0 & 0 & 1 \\ 0 & 0 & 0 \\ 1 & 0 & 0 \end{pmatrix} \frac{\partial}{\partial y} \quad .$$

The eigenvalues of $P_0(i\omega')$ are

$$\kappa_1 = 0 \quad , \quad \kappa_{2,3} = \pm\, i \sqrt{\omega_1'^2 + \omega_2'^2} = \pm\, i .$$

In [5] we have shown that our principle is also valid for hyperbolic systems. We shall discuss the results in section 3. Applications to problems in meteorology are discussed in [1], [2].

If we want to solve the above problems numerically we can procede in two ways.

1) We prepare the initial data such that the solution of the problem does not contain the fast time scale. Then we use these initial data for the difference approximation. If the difference approximation is stable then we can use standard theory to obtain error estimates. These estimates are satisfactory if the gridsize is sufficiently small such that the slow scale is resolved adequately. However, in many applications the gridsize is rather crude and therefore it is more appropriate to prepare the initial data directly for the difference approximation in a way which simulates the procedure for the differential equations.

2) Instead of preparing the initial data one could think of starting with unprepared data and use a time filter to suppress the fast scale. For linear problems this procedure is entirely satisfactory. For nonlinear problems it will only work if one can consider the fast part of the solution as a perturbation of the slow part. Probably the best procedure is to use a crude initialisation followed by time filtering. The last procedure will be discussed by G. Majda [5] in a forthcoming paper.

2. Ordinary differential equations.

We start with systems of the form

$$(2.1)\qquad \varepsilon dy/dt = A(t)y + F(t) \quad , \quad y(0) = y_0, \quad t \geq 0,$$

where $y = (y^{(1)},\dots,y^{(n)})'$, $F = (F^{(1)},\dots,F^{(n)})'$ are n dimensional vector functions, ε with $0 < \varepsilon << 1$ is a small positive constant and $A(t)$ is a smooth n×n matrix. We assume that the solutions of the homogeneous equation

$$(2.2)\qquad \varepsilon dv/dt = A(t)v$$

satisfy an estimate

$$|v(t_2)| \leq K|v(t_1)| \text{ for all } t_2, t_1 \text{ with } t_2 \geq t_1 \geq 0 .$$

Then, by Duhamel's principle, we can also estimate the solution of (2.1) and obtain

$$(2.3)\qquad |y(t)| \leq K(|y(0)| + \varepsilon^{-1}t \max_{0\leq\xi\leq t} |F(\xi)|).$$

If F has no smoothness then this estimate cannot be improved. If $F(t)$ is smooth and no turning points are present, i.e.

$$|A^{-1}(t)| \leq c,$$

then we can do much better by constructing the usual asymptotic expension. Let $\phi_0 = -A^{-1}F$ and introduce into (2.1) $y_1 = y - \phi_0$ as a new variable. Then we obtain

$$(2.4)\qquad \varepsilon dy_1/dt = A(t)y_1 + \varepsilon F_1(t), \quad y_1(0) = y_0 - \phi_0, \quad F_1 = -d\phi_0/dt.$$

(2.4) is of the same form as (2.1). Therefore we can repeat the process. After p times we obtain

$$y = \sum_{\nu=0}^{p-1} \varepsilon^\nu \phi_\nu(t) + y_p(t),$$

where the $\phi_\nu(t)$ are smooth functions (they are composed of F, A, A^{-1} and their derivatives) and $y_p(t)$ is the solution of

(2.5) $$\varepsilon dy_p/dt = A(t)y_p(t) + \varepsilon^p F_p(t),$$
$$y_p(0) = y_0 - \sum_{\nu=0}^{p-1} \varepsilon^\nu \phi_\nu(0).$$

Nowe choose the initial data y_0 by

(2.6) $$y_0 = \sum_{\nu=0}^{p-1} \varepsilon^\nu \phi_\nu(0) \quad .$$

Then it follows from (2.3) that

$$|y_p(t)| \le \text{const. } \varepsilon^{p-1} t,$$

and therefore, using (2.5) to express the derivative $d^\nu y_p/dt^\nu$ by $y_p(t)$, also

$$|d^\nu y_p(t)/dt^\nu| \le \text{const. } \varepsilon^{p-1-\nu}(t+1) \quad .$$

Thus by choosing the initial conditions according to (2.6) we have constructed a solution which has $p-1$ derivatives bounded independently of ε .

We can now prove that our principle (1.3) holds.

Theorem 2.1. Let $y(t)$ be the solution we have just constructed. Let w be another solution of (2.1) which for $t=0$ has $p-1$ derivatives bounded independently of ε , i.e.

$$|d^\nu w(0)/dt^\nu| \le c, \quad \nu=0,1,2,\dots,p-1.$$

For sufficiently small ε there is a constant K_2 such that

$$\sup_t |d^\nu(y-w)/dt^\nu| \le K_2 \varepsilon^{p-\nu-1}, \quad \nu=0,1,2,\dots,p-1.$$

Thus $w(t)$ has $p-1$ derivatives bounded independently of ε . Furthermore all solutions with this property differ from each other only by terms of order $O(\varepsilon^{p-1})$.

Proof. $v = y - w$ is the solution of the homogeneous equation

(2.7) $$\varepsilon dv/dt = Av.$$

If v has one bounded derivative at $t=0$ then

$$v(0) = \varepsilon A^{-1} dv(0)/dt = O(\varepsilon) .$$

Differentiating (2.7) it follows that $v(0) = O(\varepsilon^{p-1})$. Then the theorem follows from (2.3).

Now consider systems of the form

$$\frac{d}{dt}\begin{pmatrix} y^I \\ y^{II} \end{pmatrix} = \begin{pmatrix} \frac{1}{\varepsilon} A_{11} & \frac{1}{\varepsilon} A_{12} \\ A_{21} & A_{22} \end{pmatrix} \begin{pmatrix} y^I \\ y^{II} \end{pmatrix} + \begin{pmatrix} \frac{1}{\varepsilon} F^I \\ F^{II} \end{pmatrix} , \tag{2.8}$$

where

$$|A_{11}{}^{-1}| \le c.$$

Introducing new variables

$$y_1{}^I = y^I - A_{11}{}^{-1} A_{12} y^{II} - A_{11}{}^{-1} F^I \quad , \quad y_1{}^{II} = y^{II} \quad ,$$

we obtain a system of the form

$$\frac{d}{dt}\begin{pmatrix} y_1{}^I \\ y^{II} \end{pmatrix} = \begin{pmatrix} \frac{1}{\varepsilon} \tilde{A}_{11} & \tilde{A}_{12} \\ \tilde{A}_{21} & \tilde{A}_{22} \end{pmatrix} \begin{pmatrix} y_2{}^I \\ y^{II} \end{pmatrix} + \begin{pmatrix} \tilde{F}^I \\ \tilde{F}^{II} \end{pmatrix}$$

which is of the same form as (2.8). Therefore we can apply the transformation process once more. After p times we obtain

$$\frac{d}{dt}\begin{pmatrix} y_p{}^I \\ y^{II} \end{pmatrix} = \begin{pmatrix} \frac{1}{\varepsilon} \tilde{\tilde{A}}_{11} & \varepsilon^{p-1} \tilde{\tilde{A}}_{12} \\ \tilde{\tilde{A}}_{21} & \tilde{\tilde{A}}_{22} \end{pmatrix} \begin{pmatrix} y_p{}^I \\ y^{II} \end{pmatrix} + \begin{pmatrix} \varepsilon^{p-1} \tilde{\tilde{F}}^I \\ \tilde{\tilde{F}}^{II} \end{pmatrix} .$$

Neglecting terms of order $O(\varepsilon^{p-1})$ we obtain

$$\varepsilon dy_p{}^I/dt = \tilde{\tilde{A}}_{11} y_p{}^I \quad ,$$

$$dy^{II}/dt = \tilde{\tilde{A}}_{22} y^{II} + \tilde{\tilde{A}}_{21} y_p{}^I + \tilde{\tilde{F}}_2 \quad ,$$

and we have seperated the different time scales completely. Therefore our principle holds and by theorem 2.1 the solution of the system (2.8) has $p-1$ derivatives bounded independently of ε if and only if $y_p(0) = O(\varepsilon^{p-1})$.

The advantage of our principle is that it is invariant under transformation of the dependent variable. Therefore the system does not need to have the form (2.8). Instead we can consider general systems (2.1). Essentially the only assumption we have to make is, that the eigenvalues κ of $A(t)$ split into groups of type (1.8). We can also consider non-linear systems. In this case the linearized system must have the above properties. Details can be found in [4].

3. Partial differential equations.

In this section we consider hyperbolic systems (1.6) where $P(x,t,\partial/\partial x)$ is of the form (1.9). If we want to solve the Cauchy problem and the coefficients of P_0 do not depend on x,t then it is easy to prove that our principle is valid. We have

Theorem 3.1. Consider the Cauchy problem for the system (1.6). Assume that the coefficients of P_0 do not depend on x,t. Then our principle holds, i.e. in any finite time interval the solution and its derivatives can be estimated independently of ε by $u(x,0)$, $F(x,t)$ and their derivatives.

Proof. It is well known that the solutions of the homogeneous system

$$\partial v/\partial t = \left(\frac{1}{\varepsilon} P_0 + P_1\right)v \tag{3.1}$$

satisfy the estimate

$$||v(x,t_2)|| \leq ||v(x,t_1)||, \quad t_2 \geq t_1 ,$$

where $||\cdot||$ denotes the usual L_2-norm. Using Duhamel's principle we can therefore also estimate the solutions of the inhomogeneous problem.

Let $\dot{u}$ denote any of the first derivatives of u. Then $\dot{u}$ is the solution of

$$\begin{aligned} \partial\dot{u}/\partial t &= P\dot{u} + \dot{P}u + \dot{F} = \\ &\left(\frac{1}{\varepsilon} P_0 + P_1\right)\dot{u} + \left(\frac{1}{\varepsilon} \dot{P}_0 + \dot{P}_1\right)u + \dot{F} . \end{aligned} \tag{3.2}$$

Let $w = (\partial U/\partial t, \partial u/\partial x_1, \ldots, \partial u/\partial x_s)'$, $G = (\partial F/\partial t, \partial F/\partial x_1, \ldots, \partial F/\partial x_s)'$ denote the vector formed by the derivatives of u and F. (3.2) implies that w is the solution of

$$\partial w/\partial t = Pw + G + Rw . \tag{3.3}$$

By assumption $\dot{P}_0 \equiv 0$ and therefore

$$||Rw|| \leq \gamma ||w|| \ , \quad \gamma \text{ constant independently of } \varepsilon \ .$$

Thus, by Duhamel's principle, we can estimate $||w(x,t)||$ in terms of $\sup_t ||G||$ and $||w(x,0)||$. Repeating the process we obtain the estimates for the higher derivatives.

The proof breaks down if the coefficients of P_0 depend on x,t or if we want to solve the initial boundary value problem in a finite domain Ω . The reason is that $||R(w)|| = O(\varepsilon^{-1})$. However, the principle is still valid. The proof depends on a fundamental lemma, proved in [5], which shows that one can transform P_0 to a special normal form. Assume, for example, that the symbol $i \sum \tilde{A}_j w_j'$ in assumption 1.1 has exactly two eigenvalues $\kappa \neq 0$ then we can assume that in P_0 in three space dimensions has the form

$$P_0 u = \begin{pmatrix} a_1 & a^* \\ a & 0 \end{pmatrix} \frac{\partial u}{\partial x} + \begin{pmatrix} b_1 & b^* \\ b & 0 \end{pmatrix} \frac{\partial u}{\partial y} + \begin{pmatrix} c_1 & c^* \\ c & 0 \end{pmatrix} \frac{\partial u}{\partial z} + \begin{pmatrix} d_1 & d^* \\ d & 0 \end{pmatrix} u$$

where $a^* = (a_2, \ldots, a_n)$ and b^*, c^*, d^* are of the same form. If P_0 is of the above form then we can use the differential equations (1.6) to estimate $u^{(1)}$ and its first derivatives in terms of the other variables and their derivatives multiplied by ε. This gives a satisfactory estimate of the lower order terms in (3.3). (For details see [5]).

Finally we consider the Cauchy problem for the shallow water equations (1.10) and derive the relations the initial data must satisfy such that the solution has a number of derivatives bounded independently $\varepsilon = (gH)^{-1/2}$. The full nonlinear equations are considered in [1]. We assume that for $t = 0$ u, v, h are smooth functions of x,y. Therefore we need only to consider the time derivatives. The solutions of (1.10) have one bounded time derivative if and only if

$$h = \frac{1}{\sqrt{gH}} h_1 \ , \quad u_x + v_y = \frac{1}{\sqrt{gH}} d, \quad u_x = \partial u/\partial x, u_y = \partial u/\partial y \ , \tag{3.4}$$

where h_1, d are smooth functions. We consider now the second derivatives. Introducing (3.4) into (1.10) gives us

$$\begin{aligned} h_{1t} &= \sqrt{gH}\, d + U h_{1x} + V h_{1y}, \\ u_t &= h_{1x} + U u_x + V h_y, \\ v_t &= h_{1y} + U v_x + V v_y, \end{aligned}$$

and therefore by differentiation,

$$h_{tt} = \frac{1}{\sqrt{gH}} h_{1tt} = d_t + \frac{1}{\sqrt{gH}} (U\, h_{1xt} + V\, h_{1yt}) =$$

$$d_t + U\, d_x + V\, d_y + \frac{1}{\sqrt{gH}} (U^2 h_{1xx} + 2UVh_{1xy} + V^2 h_{1yy}),$$

$$d_t = \sqrt{gH}(u_{xt} + v_{yt}) = \sqrt{gH}(h_{1xx} + h_{1yy}) + \sqrt{gH}(Ud_x + Vd_y) \ ,$$

$$u_{tt} = h_{1xt} + U\, u_{xt} + V\, u_{yt} =$$

$$= \sqrt{gH}\, d_x + Uh_{1xx} + Vh_{1xy} + U(h_{1xx} + Uu_{xx} + Vu_{xy}) +$$

$$V(h_{1yy} + Ux_{xy} + Vv_{yy}),$$

$$v_{tt} = h_{1yt} + U\, v_{xt} + V\, v_{yt} =$$

$$\sqrt{gH}\, dy + U\, h_{1xy} + V\, h_{1yy} + U(h_{1xy} + Uv_{xx} + Vu_{xy}) +$$

$$+ V(h_{1yy} + Uv_{xy} + Vv_{yy}) \ .$$

Therefore the second derivatives are bounded if and only if

$$d = \frac{1}{\sqrt{gH}} d_1 \quad , \quad h_1 = \frac{1}{\sqrt{gH}} h_2$$

where d_1, h_2 are smooth functions. Thus

$$h = O(\frac{1}{gH}) \quad , \quad u_x + v_y = O(\frac{1}{gH}) \ .$$

Repeating this process we obtain that the solutions have p derivatives bounded independently of ε if and only if

$$d = O((gH)^{-p/2}) \quad , \quad u_x + v_y = O((gH)^{-p/2}) \ .$$

Thus in the limit we have

(3.5) $$h = 0 \quad , \quad u_x + v_y = 0 \ .$$

Furthermore by cross differentiating the last two equations (1.10) we get the vorlicity equation

$$(3.6) \qquad (u_y - v_x)_t = U(u_y - v_x)_x + V(u_y - v_x)_y \quad .$$

For $gH \to \infty$ the solutions of (1.10) with bounded derivative are the solutions of (3.5), (3.6).

References

1. Browning, G., A new system of equations for numerical weather forecasting, to appear.

2. Browning, G., Kasahara, A., Kreiss, H.O., Shallow water equations with orography, to appear.

3. Kreiss, H.O. Methods for stiff ordinary differential equations, SIAM J. Numer. Analysis, 15, 21 (1978).

4. Kreiss, H.O., Problems with different time scales for ordinary differential equations. Uppsala University, Dep. Computer Sciences, Report, 1977.

5. Kreiss, H.O., Problems with different time scales for partial differential equations, Uppsala University, Dep. Computer Sciences, Report 1978.

6. Majda, G., Time filter, to appear.

Supported in part by National Science Foundation, Grant ATM 76-10218 and Nasa Goddard Space Flight Center, Nasa-6FSC, Grant 5034.

Department of Computer Sciences
University of Uppsala
Sturegatan 4B 2TR
Uppsala S-752 23
Sweden

Accuracy and Resolution in the Computation of Solutions of Linear and Nonlinear Equations

Peter D. Lax

Let L be a linear, hyperbolic partial differential operator, with C^∞ coefficients, say a first order system:

$$L = \partial_t + \sum A_j \partial_j + B \; , \quad \partial_j = \frac{\partial}{\partial x_j} \; , \tag{1}$$

A_j and B being C^∞ matrix valued functions of x and t of order $m \times m$. We shall not define hyperbolicity but merely use the well-known result that for solutions of $Lu = 0$ the value of $u(t)$ depends boundedly on $u(0)$ in the L_2 norm, and consequently in all the Sobolev norms. In particular, if $u(x,0)$ is C^∞, so is $u(x,t)$.

We shall consider approximations to solutions u of $Lu = 0$ by solutions U of difference equations $L_h U = 0$. We take U to be defined at discrete times $0, h, 2h, \ldots$ on a rectangular spatial lattice; the time increment h is taken to be proportional to the meshsize of the lattice; for sake of simplicity we take the meshsize to be equal to h in all directions. Since L is of first order in t, it is natural to take L_h to be two-level, i.e. of the form

$$L_h = D_t^+ + \frac{1}{h} \sum S_j \, T^j \; , \tag{2}$$

D_t^+ being forward divided difference in t of steplength h, and T^j is translation in x by jh; the matrix coefficients S_j are C^∞ functions of x,t and h. As defined by (2), L_h is explicit in the positive t direction, i.e. given U^0 solutions U of $L_h U = 0$ are uniquely determined at $t = h$, and consequently at all positive integer

ISBN 0-12-208360-1

multiples of h, L_h is called stable forward, if in any given range of $t > 0$, i.e. for all $0 < nh \leq T$, U^n depends boundedly on U^0, uniformly for all h.

We say that L_h approximates L of order ν if for every C^∞ solution u of $Lu = 0$,

$$L_h(u) = O(h^\nu) \tag{3}$$

It is a basic result of the theory of difference approximations that if L_h is stable and approximates L of order ν, then for all C^∞ solutions u and for all h

$$|u(t) - U^N| = O(h^\nu), \qquad t = Nh, \tag{4}$$

where $U^0 = u(0)$, $L_h U = 0$. This shows that the higher the order of approximation, the smaller the error, for h small enough.

In many interesting problems we are presented with piecewise C^∞ initial data whose discontinuities occur along C^∞ surfaces. According to the theory of hyperbolic equations, solutions of $Lu = 0$ with such initial data are themselves piecewise C^∞, their discontinuities occurring across characteristic surfaces issuing from the discontinuity surfaces of the initial data. How good are difference approximations to discontinuous solutions? Consider the model equation

$$u_t + a\, u_x = 0, \tag{5}$$

and the difference equation obtained by replacing ∂_t by forward, ∂_x by backward difference quotients. The resulting equation can be written as

$$U_k^{n+1} = a\, U_{k-1}^n + (1-a)\, U_k^n. \tag{6}$$

It is easy to show that (6) approximates (5) of first order, and that (6) is stable if and only if $0 \leq a \leq 1$. Suppose the initial data are piecewise constants, say

$$u(x,0) = \begin{cases} 0 & \text{for } x < 0 \\ 1 & \text{for } x > 0 \end{cases}, \qquad U_k^0 = \begin{cases} 0 & \text{for } k < 0 \\ 1 & \text{for } k \geq 0 \end{cases}. \tag{7}$$

The solution of (5) with initial values (7) is

$$u(x,t) = \begin{cases} 0 & \text{for } x < at \\ 1 & \text{for } x > at. \end{cases} \tag{8}$$

The solution of (6) has a more complicated structure; we indicate it schematically so:

$$U_k^n \simeq \begin{cases} 0 & \text{for } k < an - W(n) \\ 1 & \text{for } k > an + W(n) \; ; \end{cases} \tag{9}$$

U_k^n changes gradually from near 0 to near 1 as k goes from an-W(n) to an+W(n). The width 2W(n) of this transition region is $O(\sqrt{n})$.

What happens when the first order scheme (6) is replaced by one of higher order? Since the accuracy of higher order schemes is due to small truncation error, and since the truncation error is of form $h^{\nu}E(u)$, where E is a differential operator of order $\nu+1$, it follows that the truncation error will be large around the discontinuity. Here we are in for a surprise; if U is the solution of $L_h U = 0$ with initial data (7), and L_h approximates L of order ν, then schematically U can be described by (9), the width 2W(n) of the transition region is however $O\left(n^{\frac{1}{\nu+1}}\right)$; i.e. the higher the order of accuracy, the narrower the transition region. The same, of course, is true for solutions with arbitrary discontinuous initial data.

At points away from the discontinuities the solution is C^{∞}, so there the truncation error is small; in these regions it is reasonable to use difference approximations of high order accuracy, except for the danger that the large truncation error at the discontinuities propagates into the smooth region. Majda and Osher, [8], have shown that indeed even in smooth regions $|u - U|$ is O(h) in general; they have further shown that this discrepancy can be reduced to $O(h^2)$ by the simple expedient of taking the initial data of U not as in (9), but by taking for the value of U^0 at the point k = 0 of discontinuity the arithmetic mean:

$$U_0^0 = \frac{1}{2}\,(u(-0,0) + u(+0,0)) \; . \tag{10}$$

Mock and Lax have shown in [8] that accuracy of order ν is regained if one defines the initial values of U as

$$U_k^0 = w_k u(kh,0) \, , \tag{11}$$

where the weights w_k are $=1$ for $|k| \geq \nu$, and are derived from the Gregory-Newton quadrature formula for $|k| < \nu$. It is indicated in [9] what is the appropriate analogue of (11) for discontinuous initial value problems for functions of several space variables. We shall not repeat the derivation in [9]; the basic idea is to look at moments of the solution; i.e. weighted integrals of the form

$$M(t) = \int u(x,t)\, m(x)dx\ . \tag{12}$$

It is easy to show that even for discontinuous solutions u, the moment $M(t)$ is a C^∞ function of t, provided that the weight $m(x)$ is C^∞. For if we write L in the form $L = \partial_t - G$, where G is a linear differential operator in the x variables then $L\, u = u_t - Gu = 0$ implies that

$$\partial_t^n u = G^n u\ ,$$

so that

$$\partial_t^n M = \int (\partial_t^n u)\ mdx = \int (G^n u) m\ dx = \int u\ G^{*n} m\ dx$$

where G^* is the adjoint of G. This shows that M is C^n for any n. The analysis in [9] shows that if the initial values of U are chosen according to the recipe (11), then the approximate moment

$$M_h^n = h \sum U_k^n\, m_k \tag{13}$$

differs from the exact moment $M(t)$, $t = nh$, by $O(h^\nu)$.

We turn now to nonlinear hyperbolic equations in conservation form, i.e. systems of equations

$$u_t + \operatorname{div} f(u) = 0\ , \tag{14}$$

u a vector valued function of x,t, and f a vector valued function of u. For one space variable (14) reads

$$u_t + f(u)_x = 0\ . \tag{15}$$

According to the theory of hyperbolic conservation laws, see e.g., [6], solutions of systems of the form (14), (15) are in general discontinuous. The discontinuities, called shocks, need not be present in the initial values but arise spontaneously; their speed of propagations is governed by the Rankine-Hugoniot jump relation

$$s = \frac{f^j(u_-) - f^j(u_+)}{u_-^j - u_+^j} \tag{16}$$

where u^j and f^j stand for any one of the components of the vectors u and f, and $u_\pm$, $f_\pm$ denote the values of u and f on either side of the discontinuity.

We show now that, in contrast to the linear case, the moments of discontinuous solutions of (15) are not C^∞, in fact not even C^2. This can be seen by looking at solutions of single conservation laws that at $t < 0$ contain two shocks travelling with speeds s_1 and s_2, which at time $t = 0$ collide at $x = 0$ and coalesce thereafter into a single shock propagating with speed s_3; between the shocks the solution is constant:

For $t < 0$

$$u(x,t) = \begin{cases} a & \text{for} \quad x < s_1 t \\ b & \text{for} \quad s_1 t < x < s_2 t \\ c & \text{for} \quad s_2 t < x \, . \end{cases} \tag{17$_-$}$$

For $t > 0$

$$u(x,t) = \begin{cases} a & \text{for} \quad x < s_3 t \\ & \\ c & \text{for} \quad s_3 t < x \end{cases} \; . \tag{17$_+$}$$

Clearly $u(x,t)$ defined by $(17)_\pm$ is continuous, in t, at $t = 0$. In order to satisfy (15), the jump relation (16) must be satisfied at all discontinuities:

$$s_1 = \frac{f(a) - f(b)}{a-b} \, , \qquad s_2 = \frac{f(b) - f(c)}{b-c}$$
$$s_3 = \frac{f(a) - f(c)}{a-c} \; . \tag{18}$$

Let m be any C^∞ function, and define M by (12); then, using (15) and integrating by parts

$$M_t = \int u_t \, m \, dx = -\int f_x \, m \, dx = \int f m_x \, dx \; .$$

Using the definition $(17)_{\pm}$, for $t < 0$

$$M_t = \int_{-\infty}^{s_1 t} f(a)m_x dx + \int_{s_1 t}^{s_2 t} f(b)m_x dx + \int_{s_2 t}^{\infty} f(c)m_x dx = \tag{19$_-$}$$

$$= f(a)m(s_1 t) + f(b)[m(s_2 t) - m(s_1 t)] - f(c)m(s_2 t).$$

Similarly, for $t > 0$

$$M_t = f(a)m(s_3 t) - f(c)m(s_3 t) . \tag{19$_+$}$$

Using (18) we can verify that M_t is continuous at $t = 0$. Differentiating $(19)_{\pm}$ and setting $t = 0$ we get

$$M_{tt}(-0) = s_1 f(a) + (s_2 - s_1)f(b) - s_2 f(c)$$

$$M_{tt}(+0) = s_3 f(a) - s_3 f(c) ;$$

where we assumed that $m_x(0) = 1$. It is easy to verify that, in general, $M_{tt}(-0) \neq M_{tt}(+0)$, even under the restriction (18). This shows that there are intrinsic difficulties in constructing difference schemes that would even yield moments of discontinuous solutions of nonlinear conservation laws with order of accuracy higher than first. An analysis of other difficulties, and a possible partial cure, are contained in a forthcoming article of Michael Mock.

That it is more difficult to construct accurate approximations of discontinuous solutions of nonlinear equations than of linear equations is hardly surprising. We show now that in some respects it is easier to construct them. Replace in equation (15) u_t by forward and f_x by backward difference quotients; we get

$$U_k^{n+1} = U_k^n + f(U_{k-1}^n) - f(U_k^n) . \tag{20}$$

It is easy to show that this difference scheme is stable if $0 \leq \frac{df}{du} \leq 1$. It is strongly indicated by numerical experiments, and has been shown by Jennings rigorously in [5], that if f is concave then the solution of (20) with initial values (7) is for h small described by

$$U_k^n \simeq w((k-sn)h), \tag{21}$$

where $w(x)$ is a function that tends, exponentially, to 0 as $x \to -\infty$, to 1 as $x \to +\infty$,

$$s = f(0) - f(1)$$

is the shock speed of the exact solution of (15) with initial values (7):

$$u(x,t) = \begin{cases} 0 & \text{for} \quad x < st \\ 1 & \text{for} \quad x > st \end{cases} . \tag{22}$$

The result of Jennings is in fact considerably more general than this. The important fact is that (21) is a far better approximation to (22) than (9) is to (8); the width of the transition region in (21) from near 0 to near 1 is $O(1)$, whereas in (9) that width is $O(n^{\frac{1}{\nu+1}})$! We give now a possible theoretical explanation why, as evidenced above, it might be easier to compute solutions of nonlinear equations than of linear ones. Consider a set D of initial data of interest; typically such a set might be a unit ball with respect to some Sobolev norm; this norm defines a distance in D. We denote by S the set of solutions at time t of $Lu = 0$ whose initial values belong to D. As remarked at the beginning, the mapping from D to S is bounded with respect to the norm and, since t is reversible, so is its inverse; in many important cases D and S are isometric. Denote now by D_h the projection of data in D onto data defined on a discrete lattice of mesh width h, e.g. by the formula

$$U_j = \frac{1}{h} \int_{(j-1/2)h}^{(j+1/2)h} u(x)\,dx \tag{23}$$

is one space dimension. (23) carries a ball in Sobolev space onto a ball in discrete Sobolev space defined for functions on the mesh. Define S_h to be the set filled out by U_h^n, $nh = t$, where $L_h U = 0$ and U^0 belongs to D_h. To study how good an approximation solutions of $L_h U = 0$ are to solutions of $Lu = 0$ with initial data in D, we have to obtain a uniform estimate of the distance of corresponding points of S and S_h. Before comparing them we apply the mapping (23) to S, obtaining the set S^h of functions on the mesh. The <u>approximation error</u> is defined by

$$\operatorname*{Max}_{D_h} |s^h - s_h| = \delta \tag{24}$$

where s^h is the projection via (23) of the exact solution corresponding to the initial data u_0 in D, and s_h is the approximate solution with initial data obtained from u_0 by the projection (23).

We show now how to obtain lower bounds for the approximation error δ by using some notions from information theory. We recall the definition of <u>ε-capacity</u> $C(M,\varepsilon)$ of a set M in a metric space:

$C(M,\varepsilon)$ = largest number of points in M whose distance from each other is $\geq \varepsilon$.

A related notion is the <u>ε-entropy</u> $E(M,\varepsilon)$ of M:

$E(M,\varepsilon)$ = smallest number of ε-balls that cover M; the centers of the balls need not belong to M but may lie in a metric extension of M.

Both ε-capacity and ε-entropy measure the amount of information contained in M.

<u>Theorem:</u> Denote by δ the approximation error defined by (24). Then

$$C(S^h, 3\delta) \leq C(S_h,\delta) , \tag{25}$$

and

$$E(S^h,2\delta) \leq E(S_h,\delta) . \tag{26}$$

<u>Proof:</u> By definition of C there exist $C = C(S^h,3\delta)$ points $s^1,\dots,s^C$ in S^h such that

$$|s^i - s^j| \geq 3\delta , \quad i \neq j , \tag{27}$$

By (24) the corresponding elements $s_1,\dots,s_C$ of S_h satisfy

$$|s^j - s_j| \leq \delta . \tag{28}$$

By the triangle inequality and (27), (28),

$$\begin{aligned} |s_i - s_j| &\geq |s^i - s^j| - |s^i - s_i| - |s^j - s_j| , \\ &\geq \delta . \end{aligned}$$

Thus S_h contains $C(S^h,3\delta)$ points whose distance from each other exceeds δ; this proves (25).

Similarly, by definition of E there exist $E = E(S_h,\delta)$ points $u_1,\dots,u_E$ such that every point s_h of S_h lies within δ of one of the u_j:

$$|s_h - u_j| \leq \delta.$$

By (29), $|s^h - s_h| \le \delta$, so by the triangle inequality

$$|s^h - u_j| \le |s^h - s_h| + |s_h - u_j| \le 2\delta.$$

Thus s^h can be covered by $E(S_h,\delta)$ balls of radius 2δ; this implies (26).

It was remarked at the beginning that if L is a linear hyperbolic operator, the mapping linking $u(0)$ to $u(t)$ is bounded. Since time is reversible, the inverse mapping is likewise bounded. This shows that $C(D,\varepsilon)$ and $C(S,\varepsilon)$ are comparable quantities, as are $E(D,\varepsilon)$ and $E(S,\varepsilon)$. It follows that $C(D_h,\varepsilon)$ and $C(S^h,\varepsilon)$ are likewise comparable. The forward stability of L_h means that the mapping linking D_h to S_h is bounded. But since L_h is generally unstable backwards, it is plausible to deduce (and true generally) that $C(S_h,\varepsilon)$ is very much <u>smaller</u> than $C(D_h,\varepsilon) \approx C(S^h,\varepsilon)$. It can be shown that, in general, the higher the order of accuracy of L_h , the larger the ε-capacity (and ε-entropy) of S_h . If we have lower bounds for $C(S^h,\varepsilon)$ or $E(S^h,\varepsilon)$ and upper bounds for $C(S_h,\varepsilon)$ or $E(S_h,\varepsilon)$, then using (25) or (26) we can get a **lower** bound on the approximation error δ .

Roughly speaking we shall say that an approximation method has <u>high resolution</u> if $C(S_h,\varepsilon)$ and $C(S^h,\varepsilon)$ are comparable, and <u>low resolution</u> if $C(S_h,\varepsilon)$ is very much smaller than $C(S^h,\varepsilon)$. We have shown above that a method with low resolution cannot be very accurate; the converse does not follow, i.e. a method with high resolution need not be highly accurate. But at least it furnishes approximations that contain enough information from which a better approximation may be extracted by a post-processing, hopefully at not too high an expense. Even if that isn't so, a method with high resolution is more likely to preserve qualitative features of solutions, such as number of maxima and minima, which in some cases is all we want to know.

We turn now to nonlinear conservation laws; here time is decidedly not reversible; on the contrary, here the mapping relating initial values to values at time $t > 0$ is a <u>compact</u> mapping. For single conservation laws, with f concave or convex, this follows from the explicit formula for solutions given in [5]; for 2×2 systems this follows from the estimates given in [3]; for general systems this compactness remains an intriguing conjecture. It follows from this conjecture that in the nonlinear case $C(S^h, \varepsilon)$ is much smaller than $C(D_h, \varepsilon)$, and therefore the construction of high resolution methods is easier than in the linear case. It is in this sense that approximating solutions of nonlinear initial value problems is easier than approximating solutions of linear ones.

We conclude by observing that Glimm's method, [2], which recently has been explored by Chorin as a practical one, see [1], is one of very high resolution, since it neither creates nor destroys waves. In [4] Glimm and Marchesin developed an accurate version of Glimm's method.

REFERENCES

1. Chorin, A., Random Choice Solution of Hyperbolic Systems, J. Comp. Physics, Vol. 22, 1976, 517-533.
2. Glimm, J., Solutions in the large for nonlinear hyperbolic systems of equations, Comm. Pure Appl. Math. 18 , 1965 , 697-715.
3. Glimm, J., and Lax, P. D., Decay of solutions of nonlinear hyperbolic conservation laws, Mem. Amer. Math. Soc., 101, 1970.
4. Glimm, J., and Marchesin, D., A random numerical scheme for one dimensional fluid flow with high order of accuracy, preprint to appear.
5. Jennings, G., Discrete travelling waves, Comm. Pure Appl. Math., 26, 1973, 25-37.
6. Lax, P. D., Weak solutions of nonlinear hyperbolic equations and their numerical computation, Comm. Pure Appl. Math., 7, 1954, 159-193.
7. Lax, P. D., Hyperbolic systems of conservation laws and the mathematical theory of shock waves, Vol. 11, Regional Conference Series in Appl. Math, 1973, SIAM, Philadelphia.
8. Majda, A. and Osher, S., Propagation of Error into Regions of Smoothness for Accurate Difference Approximations to Hyperbolic Equations, Comm. Pure Appl. Math., Vol. XXX, 1977, 671-706.

9. Mock, M. and Lax, P. D., The Computation of Discontinuous Solutions of Linear Hyperbolic Equations, to appear.

This work was supported in part by the U. S. Department of Energy Contract EY-76-C-02-307700 at the Courant Mathematics and Computing Laboratory, New York University.

Courant Institute of
Mathematical Sciences
New York University
New York, NY 10012

Finite Element Approximations to the One Dimensional Stefan Problem

J. A. Nitsche

0. Introduction

The mathematical formulation of many problems arising in practice leads to boundary value problems for partial differential equations - especially of parabolic type - with the feature that the boundary is not prescribed in advance but depends on certain properties of the solution itself. Probably the oldest such free boundary problem is due to Stefan (1889). In one space dimension, it may be described as follows: In a domain

$$\Omega := \{(y,\tau) \mid \tau > 0,\ 0 < y < s(\tau)\},$$

a function u is sought as the solution of the heat equation

$$u_\tau - u_{yy} = f \quad \text{in } \Omega .$$

The initial temperature $u(y,0)$ as well as the flux $u_y(0,t)$ are prescribed. The free boundary $y = s(\tau)$ is defined by the condition

$$u(s(\tau),\tau) = 0$$

on the one hand and the additional condition

$$s_\tau + u_y(s(\tau),\tau) = 0$$

on the other. The melting (or freezing) of an ice block is one of the physical interpretations.

Strongly connected with the Stefan problem is the problem of oxygen diffusion. Then the free boundary is defined by the two conditions

ISBN 0-12-208360-1

$$u(s(\tau),\tau) = 0 \ ,$$
$$u_y(s(\tau),\tau) = 0 \ .$$

The time-derivative of a solution of the second problem is a solution of the first and correspondingly by integrating a solution of the first problem with respect to time this function is a solution of the second type.

Free boundary problems have found increasing interest within the last years. The connection with variational inequalities was an additional stimulus.

In part B of the bibliography, a number of papers concerning the two problems mentioned is listed. We hope that it is representative. But we mention that the literature on free boundary problems is not covered thoroughly, e.g., the papers dealing with the 'flow through porous media' are mostly omitted.

The classical approach to solving partial differential equations is by the use of finite differences. The following papers listed in part B of the bibliography take this approach to the Stefan problem: Atthey (1974), (1975), Baiocchi-Pozzi (1977), Berger (1976), Berger-Ciment-Rogers (1975), Budak-Goldman-Uspenskii (1966), Budak-Vasilev-Uspenskii (1965), Ciment-Guenther (1974), Crank (1975), Crank - Gupta (1972a), (1972b), Douglas-Gallie (1955), Ehrlich (1958), Fasano-Primicerio (1973), Ferris (1975), Fox (1975), Gaipova (1968), Hansen-Hougaard (1974), Höhn (1978), Landau (1950), Lotkin (1960/61), Meyer (1973), Trenck (1959).

Quite often only a discretization in time is used. In this way existence and uniqueness theorems are won on the one hand and semi-discrete Galerkin-procedures are derived on the other hand. We refer to the papers of part B: Bachelis-Melamed-Shlyaiffer (1969), Baiocci-Pozzi (1976), George-Damle (1975), Jerome (1976), Kotlov (1973), Meyer (1970), (1971), (1973), (1977a), (1977b), (1978), Millinazzo-Bluman (1975), Sachs (1975), Sackett (1971), Vasilev (1968), Ventsel (1960).

The characterization of the solution of a free boundary problem by means of variational inequalities leads to a third class of numerical methods. We refer to the papers of part B: Berger (1976), Bonnerot-Jamet (1974), (1977), Brezzi-Sacchi (1976), Ciavaldini (1975), Comini-Del Guidice-Lewis-Zienkiewicz

(1974), Crank-Gupta (1972), Hunt-Nassif (1973), Wellford-Ayer (1977).

The approach presented in this paper is based on the transformation of the Stefan problem in one space dimension to an initial-boundary value problem for the heat-equation in a fixed domain. Of course, the problem is then non-linear, see Friedman (1976). The finite element approximation adopted here is the standard Galerkin method continuous in time with the modifications according to the nonlinearity. In this paper only the 'regular case' is discussed. This means the error analysis is based on the assumption that the solution is sufficiently smooth. If the initial data fulfill certain compatibility conditions, see e.g. Friedman-Jensen (1977), this assumption holds. The finite element method gives optimal order of convergence. Time discretization is not discussed here. In the regular case, no additional difficulties arise.

1. Weak Formulation of the Stefan Problem, the Finite Element Method.

The Stefan problem to be considered is the following

Problem P_U: Given $T_0 > 0$ and $g(y)$ for $y \in I := (0,1)$ with $g'(0) = g(1) = 0$. Find $\{s(\tau), U(y,\tau)\}$ such that

$$s(\tau) > 0 \quad \text{for} \quad 0 < \tau \leq T_0, \qquad s(0) = 1, \tag{1.1}$$

$$U_{yy} - U_\tau = 0 \quad \text{in} \quad \Omega = \{(y,\tau) \mid 0 < \tau \leq T_0,\ 0 < y < s(\tau)\}, \tag{1.2}$$

$$U_y(0,\tau) = 0, \quad U(s(\tau),\tau) = 0 \quad \text{for} \quad 0 < \tau \leq T_0, \tag{1.3}$$

$$U(y,0) = g(y) \quad \text{for} \quad y \in I, \tag{1.4}$$

and in addition

$$\frac{ds}{d\tau} + U_y(s(\tau),\tau) = 0 \quad \text{for} \quad 0 < \tau \leq T_0. \tag{1.5}$$

In order to reduce this problem to one with fixed boundaries, we introduce the new space variable

$$x = s^{-1}(\tau)y. \tag{1.6}$$

The corresponding transformation of τ defined by

$$\frac{d\tau}{dt} = s^2(\tau), \quad \tau(0) = 0 \tag{1.7}$$

leads for the function $u(x,t) = U(y,\tau)$ to

Problem P_u: Find u such that

$$u_{xx} - u_t = xu_x(1,t)u_x \quad \text{in} \quad Q = \{(x,t) \mid x \in I,\ 0 < t \leq T\} \tag{1.8}$$

$$\left.\begin{array}{l} u_x(0,t) = 0 \\ u(1,t) = 0 \end{array}\right\} \quad \text{for} \quad 0 < t \leq T, \tag{1.9}$$

$$u(x,0) = g(x) \quad \text{for} \quad x \in I, \tag{1.10}$$

$$\left.\begin{array}{l} \frac{ds}{dt} = -u_x(1,t)s \\ s(0) = 1 . \end{array}\right\} \quad \text{for} \quad 0 < t \leq T, \tag{1.11}$$

Here $t = T$ corresponds to $\tau = T_0$. The original Stefan problem is now split into a nonlinear parabolic initial-boundary value problem for a fixed domain and the two ordinary differential equations (1.7) and (1.11). If the boundary condition at $x = 0$ is time-dependent, the parabolic problem and the ordinary differential equations are coupled. In this more general case the results given below also hold. But the corresponding analysis is lengthier.

Let us introduce the space

$$\dot{H}_1 = \{w \mid w \in H_1(I),\ w(0) = 0\}. \tag{1.12}$$

Then

$$v = u_x \tag{1.13}$$

belongs to $\dot{H}_1$ and, for any $v \in \dot{H}_1$, the function u defined by

$$u(x,t) = -\int_x^1 v(z,t)\,dz \tag{1.14}$$

satisfies the boundary conditions (1.9). Multiplication of (1.8) by w_x with $w \in \dot{H}_1$ and integration gives

$$\int_I \{u_{xx}w_x + u_{xt}w\}dx = u_x(1,t)\int_I x\, u_x w_x ds . \tag{1.15}$$

In this way we come to

Problem P_v. Find v such that $v(\cdot,t) \in \dot{H}_1$ and

$$(\dot{v},w) + (v',w') = v(1)(xv,w') \quad \text{for } w \in \dot{H}_1 \text{ and } 0 < t \leq T \tag{1.16}$$

<u>with the initial condition</u>

$$v(\cdot,0) = g' .$$

Here and in the following, v' and $\dot{v}$ denote differentiation with respect to x and t, and $(\cdot,\cdot)$ denotes the $L_2(I)$-scalar product. The corresponding norm is $\|\cdot\|$. Most of the time, we will suppress the dependence on t, e.g., we write $v(1) = v(1,t)$ etc. (1.16) is the weak formulation of

$$v'' - \dot{v} = v(1)(xv)' \text{ in } Q = \{(x,t) \mid x \in I,\ t \in (0,T]\} \tag{1.17}$$

$$\left.\begin{aligned} v(0) &= 0 \\ v'(1) &= v^2(1) \end{aligned}\right\} \quad \text{for } 0 < t \leq T , \tag{1.18}$$

$$v = g' \qquad \text{for } t = 0 . \tag{1.19}$$

The boundary condition (1.18_2) is equivalent to

$$U_{yy} = U_\tau = U_y^2 \qquad \text{for } y = s(\tau) \tag{1.20}$$

which is a consequence of (1.5). For higher regularity of v - and in this way of u or U - the compatibility condition

$$g''(1) = g'^2 \tag{1.21}$$

is necessary.

There are many methods for getting finite element solutions. We will study only the standard one: Let $S_h = S_h^{0.r}$ denote the continuous splines of order r with $r \geq 3$, i.e., S_h consists of continuous piecewise polynomial functions of degree < 3 for some regular subdivision π of I. Up to fixed factors, h is a lower and an upper bound of the length of the subintervals of π. Further $\dot{S}_h = \{\chi \in S_h \mid \chi(0) = 0\}$. In this way, $\dot{S}_h \subseteq \dot{H}_1$. The finite element method is

<u>Problem P_{v_h}</u>. <u>Find</u> v_h <u>such that</u> $v_h(\cdot,t) \in \dot{S}_h$ <u>and</u>

$$(\dot{v}_h,\chi) + (v_h', \chi') = v_h(1)(xv_h,\chi') \quad \underline{\text{for}}\ \chi \in S_h \ \underline{\text{and}} \quad 0 < t \leq T \tag{1.22}$$

<u>with the initial condition</u>

$$v_h(\cdot,0) = Q_h\, g' . \tag{1.23}$$

Here Q_h is an appropriate projection onto $\dot{S}_h$ to be specified later.

2. Uniqueness, Error Estimates in the Regular Case.

In this section, we assume that the solution of problem P_v is sufficiently smooth. This is what is meant by 'regular case'. Then the error estimates are optimal with respect to the order of powers of h. The analysis given below could be refined to deal with reduced smoothness of the solution and the corresponding reduced order of convergence.

Since S_h is finite dimensional, problem P_{v_h} results in a finite system of ordinary differential equations. Therefore v_h always exists in a certain interval $(0,\bar{t})$. Here, $\bar{t}$ may depend on g' but not on $\dot{S}_h$. We will adopt the formulation "an estimate etc. holds locally" if there is a $\bar{t}$ depending only on the data, i.e., on g or v, such that the estimate is valid for $t \in (0,\bar{t})$.

Theorem 1: Problem P_{v_h} has a solution locally.

Proof: (1.22) with $\chi = v_h$ gives

$$\frac{d}{dt}\frac{1}{2}\|v_h\|^2 + \|v_h'\|^2 = v_h(1)(xv_h, v_h') . \tag{2.1}$$

Because of $v_h(0) = 0$, we have

$$\begin{aligned} v_h^2(1) &= 2(v_h, v_h') \\ &\leq 2\|v_h\|\|v_h'\| \end{aligned} \tag{2.2}$$

and

$$(xv_h, v_h') \leq \|v_h\|\,\|v_h'\| . \tag{2.3}$$

Therefore we get

$$\begin{aligned} \frac{d}{dt}\|v_h\|^2 + 2\|v_h'\|^2 &\leq 2\sqrt{2}\|v_h\|^{3/2}\|v_h'\|^{3/2} \\ &\leq 2\|v_h'\|^2 + c\|v_h\|^4 . \end{aligned} \tag{2.4}$$

Here and in the following c, $c(\|g'\|), \ldots$ denote constants which may be different at different places. From this, standard arguments give the inequality

$$\|v_h(t)\|^2 \leq \frac{\|Q_h g'\|^2}{1 - c\|Q_h g'\|^2 t} . \tag{2.5}$$

The operators Q_h to be used will be bounded uniformly,

$$\|Q_h g'\| \leq c(g) . \tag{2.6}$$

Then v_h exists uniformly in h for $t < (c\, c(g))^{-1}$. |||

The uniqueness can be shown globally:

<u>Theorem 2:</u> <u>For any</u> $K > 0$ <u>fixed, there is at most one solution</u> v_h <u>with</u>

$$v_h(\cdot,t) \in B_K = \{w | w \in L_\infty(I), \|w\|_{L_\infty} \leq K\}. \tag{2.7}$$

<u>Proof:</u> The difference $\Phi = v_h^1 - v_h^2$ of two solutions solves

$$(\dot{\Phi},\chi) + (\Phi',\chi') = v_h^1(1)(x\Phi,\chi') + \Phi(1)(xv_h^2,\chi') \tag{2.8}$$

$$\text{for } \chi \in \dot{S}_h \text{ and } t > 0 ,$$

$$(\cdot,0) = 0 . \tag{2.9}$$

The choice $\chi = \Phi$ gives

$$\frac{d}{dt}\frac{1}{2}\|\Phi\|^2 + \|\Phi'\|^2 \leq K\{\|\Phi\|\|\Phi'\| + |\Phi(1)|\|\Phi'\|\} \tag{2.10}$$

This implies

$$\frac{d}{dt}\|\phi\|^2 \leq c(K)\|\Phi\|^2 , \tag{2.11}$$

and $\Phi \equiv 0$ is the consequence. |||

In order to simplify the presentation, we introduce the trilinear form

$$b(\xi,\eta,\zeta) = \frac{1}{2}\xi(1)(x\eta,\zeta') + \frac{1}{2}\eta(1)(x\xi,\zeta') . \tag{2.12}$$

Both $w = v$ and $w = v_h$ satisfy

$$(\dot{w},\chi) + (w',\chi') = b(w,w,\chi) \quad \text{for } \chi \in \dot{S}_h . \tag{2.13}$$

For the error $e = e_h = v-v_h$, we get by subtraction the defining relation

$$(\dot{e},\chi) + (e',\chi') - 2b(v,e,\chi) = -b(e,e,\chi) . \tag{2.14}$$

Since v is fixed,

$$a(\xi,\eta) = (\xi',\eta') - 2b(v,\xi,\eta) \tag{2.15}$$

is a bilinear form in ξ,η. In the appendix we will prove

<u>Lemma 1:</u> (i) a <u>is bounded in</u> $\dot{H}_1$, <u>i.e.</u>

$$|a(\xi,\eta)| \leq M\|\xi'\|\|\eta'\| , \tag{2.16}$$

(ii) a <u>is coercive in</u> $\dot{H}_1$, <u>i.e.</u>

$$a(\xi,\xi) \geq m\|\xi'\|^2 - \Lambda\|\xi\|^2 . \tag{2.17}$$

Here $m > 0$, M, Λ depend only on $\|v\|_{L_\infty(L_\infty)}$. Then

$$a_\Lambda(\xi,\eta) = a(\xi,\eta) + \Lambda(\xi,\eta) \tag{2.18}$$

is positive in $\dot{H}_1$ and (2.14) can be rewritten

$$(\dot{e},\chi) + a_\Lambda(e,\chi) = \Lambda(e,\chi) - b(e,e,\chi) \ . \tag{2.19}$$

In order to derive error estimates, we need a splitting of e. Let $\tilde{v}_h = Q_h v \in \dot{S}_h$ be the Galerkin approximation to v with respect to the form a_Λ :

$$a_\Lambda(v-\tilde{v}_h,\chi) = 0 \qquad \text{for} \qquad \chi \in \dot{S}_h \ . \tag{2.20}$$

In this way the operator Q_h is defined. Then we put

$$\begin{aligned} e &= (v-\tilde{v}_h) - (v_h-\tilde{v}_h) \\ &= \varepsilon - \Phi \ . \end{aligned} \tag{2.21}$$

The correction term Φ is in $\dot{S}_h$. We get from (2.19)

$$(\dot{\Phi},\chi) + a_\Lambda(\Phi,\chi) = \Lambda(\Phi,\chi) + (\dot{\varepsilon},\chi) - \Lambda(\varepsilon,\chi) + e(1)(xe,\chi') . \tag{2.22}$$

In view of Theorem 2, our aim is to show the existence of a finite element solution in the neighbourhood of v . In order to achieve this, we replace in the quadratic term $e(1)$ by $E(1)$, for some function E . Then

$$\begin{aligned} (\dot{\Phi},\chi) + a_\Lambda(\Phi,\chi) = \Lambda(\Phi,\chi) - E(1)(x\Phi,\chi') &+ \\ + (\dot{\varepsilon},\chi) - \Lambda(\varepsilon,\chi) &+ E(1)(x\varepsilon,\chi') \end{aligned} \tag{2.23}$$

is a linear problem. Therefore, for any $E(1) = E(1,t)$, there exists a solution Φ of (2.23) with

$$\Phi(\cdot,0) = 0 \ . \tag{2.24}$$

From Φ , we obtain $e = \varepsilon - \Phi$ by (2.21), with e now depending on E . We will show that there is an E with $e = E$.

<u>Lemma 2:</u> <u>Let</u> $\cdot E > 0$ <u>be fixed and</u> $E(1) = E(1,t)$ <u>be measurable and bounded by</u> 1 . <u>Then</u>

$$\begin{aligned} \|\Phi\|_{L_\infty(L_2)} \le c\{\|\varepsilon\|_{L_2(H_{-1})} &+ \|\dot{\varepsilon}\|_{L_2(H_{-1})} \\ &+ \|E\|_{L_\infty(L_\infty)}\|\varepsilon\|_{L_2(L_2)}\} . \end{aligned} \tag{2.25}$$

$L_\infty(L_\infty)$ is the abbreviation for $L_\infty(0,\bar{t};L_\infty)$ and the norm in H_{-1} is defined by

$$\| \xi \|_{-1} = \sup\{ (\xi,\eta) \mid \eta \in \dot{H}_1,\ \| \eta' \| \le 1 \}. \tag{2.26}$$

<u>Proof:</u> Using Lemma 1, we get from (2.23) with $\chi = \Phi$

$$\frac{1}{2}\frac{d}{dt} \| \Phi \|^2 + m \| \Phi' \|^2 \le \{ (\Lambda+1) \| \Phi \| + \| \dot{\varepsilon} \|_{-1} + \Lambda \| \varepsilon \|_{-1} + \| E \|_{L_\infty} \| \varepsilon \| \} \| \Phi' \| \tag{2.27}$$

and further

$$\frac{d}{dt} \| \Phi \|^2 \le c\{ \| \Phi \|^2 + \| \dot{\varepsilon} \|^2_{-1} + \| \varepsilon \|^2_{-1} + \|E\|^2_{L_\infty} \| \varepsilon \|^2 \} \tag{2.28}$$

Since $\Phi(\cdot,0)$, Gronwall's Lemma gives Lemma 2. |||

Inequality (2.25) gives for the $L_\infty(L_2)$-norm of e the bound

$$\| e \|_{L_\infty(L_2)} \le c\{ \| \varepsilon \|_{L_\infty(L_2)} + \| \dot{\varepsilon} \|_{L_2(H_{-1})} + \| E \|_{L_\infty(L_\infty)} \| \varepsilon \|_{L_2(L_2)} \}. \tag{2.29}$$

On the other hand, applying the inverse property

$$\| \chi \|_{L_\infty} \le c\, h^{-1/2} \| \chi \|_{L_2} \quad \text{for} \quad \chi \in S_h , \tag{2.30}$$

we get from (2.25) also

$$\| e \|_{L_\infty(L_\infty)} \le \| \varepsilon \|_{L_\infty(L_\infty)} + c\, h^{-1/2} \{ \| \varepsilon \|_{L_2(H_{-1})} + \| \dot{\varepsilon} \|_{L_2(H_{-1})} + \| E \|_{L_\infty(L_\infty)} \| \varepsilon \|_{L_2(L_2)} \}. \tag{2.31}$$

In the appendix we will prove

<u>Lemma 3:</u> <u>Assume</u> $r \ge 3$, <u>i.e. the approximation spaces are at least quadratic splines. Let the exact solution</u> v <u>of problem</u> P_v <u>be sufficiently smooth. Then for any time</u> t <u>fixed</u>

$$\| \varepsilon \|_{H_k} + \| \dot{\varepsilon} \|_{H_k} \le c\, h^{r-k} \quad \text{for} \quad -1 \le k \le 1 , \tag{2.32}$$

$$\| \varepsilon \|_{L_\infty} \le c\, h^r . \tag{2.33}$$

Then we have for the solution e of the linearized problem (2.23)

$$\begin{aligned} \| e \|_{L_\infty(L_2)} &\le c\, h^r (1 + \| E \|_{L_\infty(L_\infty)}) , \\ \| e \|_{L_\infty(L_\infty)} &\le c (h^r + h^{r-1/2} \| E \|_{L_\infty(L_\infty)}) . \end{aligned} \tag{2.34}$$

We have $r \geq 3$. The image e of any E with

$$E \in B_1 = \{w \mid \|w\|_{L_\infty(L_\infty)} \leq 1\} \tag{2.35}$$

is contained in B_1 for $h \leq \bar{h}$ with appropriate

$$\bar{h} < 1/(2c) \; . \tag{2.36}$$

Schauder's fixpoint-theorem guarantees the existence of an E with $e = E$. Thus we have proved:

<u>Theorem 3:</u> <u>Let</u> $\bar{t} > 0$ <u>be chosen properly and assume that the solution of problem</u> P_v <u>is sufficiently smooth in</u> $I \times [0,\bar{t}]$. <u>Then there exists exactly one finite element solution</u> v_h <u>in the neighbourhood of</u> v <u>with optimal order of convergence</u>

$$\|v-v_h\|_{L_\infty(0,\bar{t};L_\infty)} = O(h^r) \; . \tag{2.37}$$

<u>Remark:</u> Using Lemma 3, we have for a function g' sufficiently smooth

$$\|Q_h g'\| \leq \|g'\| + O(h^r) \; . \tag{2.38}$$

Therefore (2.6) is valid.

The finite-element solution v_h does not give directly an approximation to the function U defined by problem P_U. An approximation u_h for the solution u of problem P_u is given by

$$u_h(x,t) = -\int_x^1 v_h(z,t)\,dz \; . \tag{2.39}$$

The transformation $(y,\tau) \Leftrightarrow (x,t)$ is approximated - see (1.7) and (1.11) - by

$$\begin{aligned} y &= s_h(t)x \; , \\ \tau &= \tau_h(t) \end{aligned} \tag{2.40}$$

with s_h, τ_h defined by

$$\begin{aligned} \dot{s}_h &= -v_h(1)s_h \quad && \text{with} \quad s_h(0) = 1 \; , \\ \dot{\tau}_h &= s_h^2 && \text{with} \quad \tau_h(0) = 0 \; . \end{aligned} \tag{2.41}$$

Then

$$U_h(y,\tau) = u_h(x,t) \tag{2.42}$$

is the approximation for u. Theorem 3 gives at once

Corollary 3: Under the assumptions of Theorem 3, for some positive t_0, the errors

$$\|s-s_h\|_{L_\infty(0,t_0)}\ ,$$
$$\|\tau-\tau_h\|_{L_\infty(0,t_0)}\ , \tag{2.43}$$
$$\|U-U_h\|_{L_\infty(0,t_0;W^1_\infty)}$$

are of order h^r.

The last norm is to be interpreted as follows:

$$\|w\|_{L_\infty(0,t_0;W^1_\infty)} = \sup_{0\le t\le t_0}\{\|w\|_{L_\infty(\hat{I}_t)} + \|w'\|_{L_\infty(\hat{I}_t)}\} \tag{2.44}$$

with

$$\hat{I}_t = (0,\mathrm{Min}(s(t),s_h(t))). \tag{2.45}$$

Remark: The function u_h (2.39) is a spline of order $r+1$. An order $r+1$ of convergence might be expected. But since the error in the transformation $(x,t) \mapsto (y,\tau)$ is of order r, this would not give an improvement.

3. Appendix: Proof of Lemmata 1 and 3.

By definition, we have

$$a(\xi,\eta) = (\xi',\eta') - 2b(v,\xi,\eta) \tag{3.1}$$

with

$$2b(v,\xi,\eta) = \xi(1)(xv,\eta') + v(1)(x\xi,\eta'). \tag{3.2}$$

Using (2.2), we get for $\xi,\eta \in \dot{H}_1$

$$2|b(v,\xi,\eta)| \le c\{\|\xi\|^{1/2}\|\xi'\|^{1/2} + \|\xi\|\}\|\eta'\|. \tag{3.3}$$

Here c depends only on $\|v\|_{L_\infty}$. Since

$$\|\xi\| \le \|\xi'\| \tag{3.4}$$

holds for $\xi \in \dot{H}_1$, (2.16) is shown with $M = 1 + 2c$.

On the other hand we get

$$a(\xi,\xi) \ge \|\xi'\|^2 - 2c\|\xi\|^{1/2}\|\xi'\|^{3/2} \tag{3.5}$$

and therefore, because of Young's inequality

$$ab \le \frac{1}{p}a^p + \frac{1}{q}b^q \tag{3.6}$$

(valid for $a,b > 0$ and $p^{-1} + q^{-1} = 1$), we have

$$a(\xi,\xi) \geq \frac{1}{4}\|\xi'\|^2 - 4c^4\|\xi\|^2 . \tag{3.7}$$

This completes the proof of Lemma 1 .

We will begin the proof of Lemma 3 by proving the error bounds (2.32) for ε . For any $w \in \dot{H}_1$, let now $\tilde{w}_h \in \dot{S}_h$ be the Galerkin approximation defined by

$$a_\Lambda(w-\tilde{w}_h,\chi) = 0 \quad \text{for } \chi \in \dot{S}_h . \tag{3.8}$$

By standard arguments - see Aziz-Babuška (1972), pp. 185-188 - the error bound

$$\|w'-\tilde{w}_h'\| \leq (1+m^{-1}(M+\Lambda)) \inf\{\|w'-\chi'\| \,|\, \chi \in \dot{S}_h\} \tag{3.9}$$

is derived. By approximation theory, the estimate

$$\|w'-\tilde{w}_h'\| \leq ch^{\rho-1}\|w^{(\rho)}\| \quad \text{for } 1 \leq \rho \leq r \tag{3.10}$$

follows, which proves (2.32) for $k = 1$.

Although the bilinear form $a(\cdot,\cdot)$ is not symmetric, the duality argument of Aubin (1967) - Nitsche (1968) can be applied. Let the second order differential operator A be defined by

$$Aw := -w'' + \Lambda w - v(1)xw' \tag{3.11}$$

with

$$\dot{H}_2 := D(A) = H_2(I) \cap \{w|w(0) = w'(1) - (xv,w') = 0\} \tag{3.12}$$

being the domain of definition. Then

$$a_\Lambda(\xi,\eta) = (\xi,A\eta) \tag{3.13}$$

for $\xi \in \dot{H}_1$ and $\eta \in \dot{H}_2$. Now because of the positivity of a_Λ there exists for any $\eta \in L_2$ a $\zeta \in \dot{H}_2$ with

$$A\zeta = \eta . \tag{3.14}$$

Moreover, A is a bounded map on $\dot{H}_2 \cap H_{k+2}$ onto H_k with bounded inverse for $k \geq 0$, i.e.

$$c_k^{-1}\|\eta\|_{H_k} \leq \|A^{-1}\eta\|_{H_{k+2}} \leq c_k\|\eta\|_{H_k} \tag{3.15}$$

with c_k depending only on $\|v\|_{L_\infty}$.

Now let

$$\varepsilon_w = w - \tilde{w}_h \tag{3.16}$$

and define $z \in \dot{H}_2$ by

$$Az = \varepsilon_w . \tag{3.17}$$

Then we have - using (3.8) with $\chi \in \dot{S}_h$ -

$$\begin{aligned} \|\varepsilon_w\|^2 &= a_\Lambda(\varepsilon_w, z) \\ &= a_\Lambda(\varepsilon_w, z-\chi) \\ &\leq c\|\varepsilon_w'\|\|z'-\chi'\| \end{aligned} \tag{3.18}$$

With χ chosen properly we get

$$\begin{aligned} \|z'-\chi'\| &\leq c\, h\|z''\| \\ &\leq c\, h\|\varepsilon_w\| \end{aligned} \tag{3.19}$$

and therefore

$$\|\varepsilon_w\|^2 \leq c\, h\|\varepsilon_w\|\|\varepsilon_w'\| \ . \tag{3.20}$$

This gives (2.32) for $k = 0$. Finally we have, for any $z \in \dot{H}_1$ and $Z \in \dot{H}_2$ satisfying the relation

$$AZ = z \ , \tag{3.21}$$

the estimate

$$\begin{aligned} |(\varepsilon_w, z)| &= |a_\Lambda(\varepsilon_w, Z-\chi)| \\ &\leq c\, \|\varepsilon_w'\|\ \|Z'-\chi'\| \\ &\leq c\, h^2\|\varepsilon_w'\|\ \|z'\| . \end{aligned} \tag{3.22}$$

In this way (2.32) is shown for $k = -1$.

Remark: In order to get the factor h^2 in the last inequality, the assumption $r \geq 3$ is necessary.

It remains to prove (2.33). We will apply an argument due to Wheeler (1973) for splines only continuous and to Douglas-Dupont-Wahlbin (1975) for splines with higher smoothness. For $w \in \dot{H}_1$, let $\hat{w}_h \in \dot{S}_h$ be the Ritz-approximation defined by

$$(w'-\hat{w}_h', \chi') = 0 \qquad \text{for} \qquad \chi \in \dot{S}_h \ . \tag{3.23}$$

Then for functions w sufficiently smooth the error estimate

$$\|w-\hat{w}_h\|_{L_\infty} = O(h^r) \tag{3.24}$$

holds. The difference

$$\Phi = \tilde{w}_h - \hat{w}_h \in \dot{S}_h \tag{3.25}$$

obeys

$$a_\Lambda(\Phi, \chi) = -2b(v, w-\hat{w}_h, \chi) + \Lambda(w-\hat{w}_h, \chi) \ . \tag{3.26}$$

The choice $\chi = \Phi$ gives

$$m \| \Phi' \|^2 \leq c\{\| w-\hat{w}_h \| + \| w-\hat{w}_h \|_{L_\infty}\} \| \Phi' \| .$$

Because of

$$\| \varepsilon_w \|_{L_\infty} \leq \| w-\hat{w}_h \|_{L_\infty} + \| \Phi \|_{L_\infty} \tag{3.27}$$

and

$$\| \Phi \|_{L_\infty} \leq \| \Phi' \| \leq c \| w-\hat{w}_h \|_{L_\infty} , \tag{3.28}$$

the inequality (2.33) is proved.

The function $w = w(\cdot,t)$ may depend on t . Then $\varepsilon_w = w-\tilde{w}_h$ is also a function of t . By differentiation of (3.8) we get

$$a_\Lambda(\dot{\varepsilon}_w,\chi) = \varepsilon_w(1)(x\dot{v},\chi') + \dot{v}(1)(x\varepsilon_w,\chi') \quad \text{for} \quad \chi \in \dot{S}_h . \tag{3.29}$$

By arguments similar to those above we get the estimates (2.32) for $\dot{\varepsilon}_w$.

References A: Literature Cited Explicitly

Aubin, J.-P. (1967): Approximation des espaces de distributions et des opérateurs différentiels. Bull. Soc. Math. France Suppl. Mém. 12.

Aziz, K. and I. Babuška (1972): The mathematical foundations of the finite element method with application to partial differential equations. Acad. Press, New York and London.

Douglas, J., Jr., T. Dupont and L. Wahlbin (1975): Optimal L_∞-error estimates for Galerkin approximations to solutions of two point boundary value problems. Math. Comp. 29 , 475-483.

Friedman, A. (1976): Analyticity of the free boundary for the Stefan problem. Arch. Rat. Mech. and Anal. 61, 97-125.

Friedman A. and R. Jensen (1977): Convexity of the free boundary in the Stefan problem and in the dam problem. Arch. Rat. Mech. and Anal. 67, 1-24.

Nitsche, J. A. (1968): Ein Kriterium für die Quasi-Optimalität des Ritzschen Verfahrens. Num. Math. 11, 346-348.

Wheeler, M. (1973): An optimal L_∞-error estimate for Galerkin approximations to solutions of two-point boundary value problems. SIAM J. Numer. Anal. 10, 914-917.

References B: Literature on Free Boundary Problems

Acker, A. (1978): Some free boundary optimization problems and their solutions. ISNM 39, 9-22.

Agrawal, H. C. (1975): Biot's variational principle for moving boundary problems. Moving boundary problems in heat flow and diffusion. J. R. Ockendon and W. R. Hodgkins eds., Clarendon Press, Oxford, 242-250.

Atthey, D. R. (1974): A finite difference scheme for melting problems. J. Inst. Math. Appl. 13, 353-366.

Atthey, D. R. (1975): A finite difference scheme for melting problems based on the method of weak solutions. Moving boundary problems in heat flow and diffusion. J. R. Ockendon and W. R. Hodgkins eds., Clarendon Press, Oxford, 182-191.

Bachelis, R. D., V. G. Melamed and D. B. Shlyaiffer (1969): The solution of the problem of Stefan type by the straight line method. USSR Comp. Math. and math Phys., 9, 3, 113-126.

Baiocchi, C. and G. A. Pozzi (1976): An evolution variational inequality related to a diffusion-absorption problem. Appl. Math. a. Optim. 2, 304-314.

Baiocchi, C. and G. A. Pozzi (1977): Error estimates and free-boundary convergence for a finite difference discretization of a parabolic variational inequality. R.A.I.R.O. Numer. Anal. 11, 315-340.

Barenblatt, G. I. and A. Yu. Ishlinskii (1962): On the impact of a viscoplastic bar on a rigid obstacle. J. Appl. Math. Mech. 26, 740-748.

Bensoussan, A. and A. Friedman (1977): Nonzero-sum stochastic differential games with stopping times and free boundary problems. Trans. AMS 231, 275-327.

Berger, E. (1976): The truncation method for the solution of a class of variational inequalities. R.A.I.R.O. Numer. Anal. 10, 29-42.

Berger, A. E., M. Ciment and J. Rogers (1975): Numerical solution of a diffusion consumption problem with a free boundary. SIAM J. Numer. Anal. 12, 646-672.

Boley, B. A. (1961): A method of heat conduction analysis of melting and solidification problems. J. Math. Phys. 40, 300-313.

Boley, B. A. (1963): Upper and lower bounds for the solution of a melting problem. Quart. J. Appl. Math. 21, 1-11.

Boley, B. A. (1964a): Upper and lower bounds in problems of melting or solidifying slabs. Quart. J. Mech. Appl. Math. 17, 253-269.

Boley, B. A. (1964b): Estimate of errors in approximate temperature and thermal stress calculation. Proceedings XI. Intern. Congress of Appl. Mech., Springer-Verlag, 586-596.

Boley, B. A. (1968): A general starting solution for melting or solidifying slabs. Int. J. Engng. 6 , 89-111.

Boley, B. A. (1970): Uniqueness in a melting slab with space- and time-dependent heating. Quart. J. Appl. Math. 27, 481-487.

Boley, B. A. (1975): The embedding technique in melting and solidification problems. Moving boundary problems in heat flow and diffusion. J. R. Ockendon and W. R. Hodgkins eds., Clarendon Press, Oxford, 150-172.

Boley, B. A. and H. P. Yagoda (1969: The starting solution for two-dimensional heat conduction problems with change of phase. Quart. J. Appl. Math. 27, 223-246.

Bonnerot, R. and P. Jamet (1974): A second order finite element method for the one-dimensional Stefan problem. Int. J. Num. Meth. in Engng. 8, 811-820.

Bonnerot, R. and P. Jamet (1977): Numerical computation of the free boundary for the two-dimensional Stefan problem by space time finite elements. J. Comp. Phys. 25, 163-181.

Brezzi, F. and G. Sacchi (1976): A finite element approximation of a variational inequality related to hydraulics. Calcolo XIII, 257-274.

Budak, B. M., N. A. Goldman and A. B. Uspenskii (1966): Difference schemes with rectification of the fronts for solving multifront Stefan problems. Soviet Math. Dokl. 7, 454-458.

Budak, B. M. and M. Z. Moskal (1969a): On the classical solution of the Stefan problem. Sov. Math. Dokl. 10, 219-223.

Budak, B. M. and M. Z. Moskal (1969b): Classical solutions of the multi-dimensional multifront Stefan problem. Sov. Math. Dokl. 10, 1043-1046.

Budak, B. M., F. P. Vasilev and A. B. Uspenskii (1965): Difference methods for solving certain boundary value problems. USSR Comput. Math. math. Phys. 5, (5), 59-76.

Caffarelli, L. A. and N. M. Riviere (1976): Smoothness and analyticity of free boundaries in variational inequalities. Ann. Scuola norm. sup. Pisa 3 (IV), 289-310.

Cannon, J. R. and J. Douglas (1967a): The Cauchy problem for the heat equation. SIAM J. Numer. Anal. 4, 317-336.

Cannon, J. R. and J. Douglas (1967b): The stability of the boundary in a Stefan problem. Ann. d. Scuola norm. sup. di Pisa 21 (III), 83-91.

Cannon, J. R., J. Douglas and C. D. Hill (1967): A multi boundary Stefan problem and the disappearance of phases. J. Math. Mech. 17, 21-33.

Cannon, J. R. and R. E. Ewing (1976): A direct numerical procedure for the Cauchy problem for the heat equation. J. Math. Anal. Appl. 56, 7-17.

Cannon, J. R. and A. Fasano (1977): A nonlinear parabolic free boundary value problem. Ann. Mat. P. Appl. 112 , 119-149.

Cannon, J. R., D. Henry and D. Kotlow (1974). Continuous differentiability of the free boundary for weak solutions of the Stefan problem. Bull. AMS 80, 45-48.

Cannon, J. R. and C. D. Hill (1967a): Existence, uniqueness, stability, and monotone dependence in a Stefan problem for the heat equation. J. Math. Mech. 17, 1-19.

Cannon, J. R. and C. D. Hill (1967b): Remarks on a Stefan problem. J. Math. Mech. 17, 433-441.

Cannon, J. R. and C. D. Hill (1968): On the infinite differentiability of the free boundary in a Stefan problem. J. Math. Anal. Appl. 22, 385-397.

Cannon, J. R. and C. D. Hill (1970): On the movement of a chemical reaction interface. Ind. Math. J. 20, 429-454.

Cannon, J. R., C. D. Hill and M. Primicerio (1970): The one-phase Stefan problem for the heat equations with boundary temperature specifications. Arch. Rat. Mech. Anal. 39, 270-274.

Cannon, J. R. and M. Primicerio (1971a): A two-phase Stefan problem with temperature boundary conditions. Ann. Mat. P. Appl. 88, 177-192.

Cannon, J. R. and M. Primicerio (1971b): A two-phase Stefan problem with flux boundary conditions. Ann. Mat. P. Appl. 88, 193-206.

O'Carroll, M. J. and H. T. Harrison (1976): A variational method for free boundary problems. MAFELAP 1975, Proceedings of the Brunel University conference of the Institute of Mathematics and its Applications, J. R. Whiteman ed., Academic Press, London, 143-150.

Chan, C. Y. (1970): Continuous dependence on the data for a Stefan problem. SIAM J. Math. Anal. 1, 282-288.

Chan, C. Y. (1971): Uniqueness of a monotone free boundary problem. SIAM J. Appl. Math. 20, 189-194.

Ciavaldini, J. F. (1975): Analyse numérique d'un problème de Stefan a deux phases par une méthode d'éléments finis. SIAM J. Numer. Anal. 12, 464-487.

Ciment, M. and R. B. Guenther (1974): Numerical solution of a free boundary value problem for parabolic equations. Appl. Analysis 4, 39-62.

Comini, G., S. Del Guidice, R. W. Lewis and O. C. Zienkiewicz (1974): Finite element solution of non-linear heat conduction problems with special reference to phase change. Int. J. Num. Meth. Engng. 8, 613-624.

Crank, J. (1957): Two methods for the numerical solution of moving boundary problems in diffusion and heat flow. Quart. J. Mech. Appl. Math. 10, 220-231.

Crank, J. (1975): Finite-difference methods. Moving boundary problems in heat flow and diffusion. J. R. Ockendon and W. R. Hodgkins eds., Clarendon Press, Oxford, 192-207.

Crank, J. and R. S. Gupta (1972a): A moving boundary problem arising from the diffusion of oxygen in absorbing tissue. J. Inst. Math. Applics. 10, 19-33.

Crank, J. and R. S. Gupta (1972b): A method for solving moving boundary problems in heat flow using cubic splines or polynomials. J. Inst. Math. Applics. 10, 296-304.

Crank, J. and R. D. Phahle (1973): Melting ice by the isotherm migration method. Bull. Inst. Math. Applics. 9 , 12-14.

Crowley, A. B. and J. R. Ockendon (1977): A Stefan problem with a non-monotone boundary. J. Inst. Math. Applics. 20, 269-281.

Cryer, C. W. (1970): On the approximate solution of free boundary problems using finite differences. J. Assoc. Comput. Mach. 17, 397-411.

Douglas, J. (1957): A uniqueness theorem for the solution of a Stefan problem. Proc. AMS 8, 402-408.

Douglas, J. and T. M. Gallie (1955): On the numerical integration of a parabolic differential equation subject to a moving boundary condition. Duke Math. J. 22, 557-571.

Duvaut, G. (1973): Résolution d'un problème de Stefan. C.R. Acad. Sci. Paris, 276, 1461-1463.

Duvaut, G. (1974): Résolution d'un problème de Stefan. New Variational Techniques in Math. Phys., C.I.M.E., Cremonese, 84-102.

Duvaut, G. (1975): The solution of a two-phase Stefan problem by a variational inequality. Moving boundary problems in heat flow and diffusion. J. R. Ockendon and W. R. Hodgkins eds., Clarendon Press, Oxford, 173-191.

Ehrlich, L. W. (1958): A numerical method of solving a heat flow problem with moving boundary. J. Assoc. Comp. Mach. 5, 161 176.

Elliott, C. M. (1978): Moving boundary problems and linear complementarity. ISNM 39, 62-73.

Evans, G. W. (1951): A note on the existence of a solution to a problem of Stefan. Quart. J. Appl. Math. 9, 185-193.

Evans, G. W., E. Isaacson and I. MacDonald (1950): Stefan-like problems. Quart. J. Appl. Math. 8, 312-319.

Evans, L. C. (1977): A free boundary problem: the flow of two immiscible fluids in a one-dimensional porous medium, I.; Ind. Math. J. 26, 915-952.

Evans, N.T.S. and A.R. Gourlay (1977): The solution of a two-dimensional time-dependent diffusion problem concerned with oxygen metabolism in tissues. J. Inst. Math. Applics. 19, 239-251.

Fasano, A. and M. Primicerio (1973): Convergence of Huber's method for heat conduction problems with change of phase. ZAMM 53, 341-348.

Fasano. A. and M. Primicerio (1977a): General free-boundary problems for the heat equation, I.; J. of Math. Anal. and Appl. 57, 694-723.

Fasano, A. and M. Primicerio (1977b): General free-boundary problems for the heat equation, III.; J. of Math. Anal. and Appl. 59, 1-14.

Ferris, D. H. (1975): Fixation of a moving boundary by means of a change of independent variable. J. R. Ockendon and W. R. Hodgkins eds., Clarendon Press, Oxford, 251-255.

Fife, P. C. and J. B. McLeod (1977): The approach of solutions of nonlinear diffusion equations to travelling front solutions. Arch. Rat. Mech. and Anal. 65, 335-377.

Fox, L. (1975): What are the best numerical methods? Moving boundary problems in heat flow and diffusion. J. R. Ockendon and W. R. Hodgkins eds., Clarendon Press, Oxford, 210-241.

Friedman, A. (1959): Free boundary problems for parabolic equations. I: Melting of solids. J. Math. Mech. 8, 499-518.

Friedman, A. (1960a): Free boundary problems for parabolic equations, II: Evaporation and condensation of a liquid drop. J. Math. Mech. 9 , 19-66.

Freidman, A. (1960b): Free boundary problems for parabolic equations. III: Dissolution of a gas bubble in liquid. J. Math. Mech. 9, 327-345.

Friedman, A. (1960c): Remarks on Stefan-type free boundary problems for parabolic equations. J. Math. Mech. 9 , 885-904.

Friedman, A. (1968a): The Stefan problem in several space variables. Trans. AMS 133, 51-87.

Friedman, A. (1968b): One dimensional Stefan problems with non-monotone free boundary. Trans. AMS 133 , 89-114.

Friedman, A. (1975): Parabolic variational inequalities in one space dimension and smoothness of the free boundary. J. Functional Anal. 18, 151-176.

Friedman, A. (1976): A class of quasi-variational inequalities II. J. Diff. Equ. 22, 379-401.

Friedman, A. and R. Jensen (1975): A parabolic quasi-variational inequality arising in hydraulics. Ann. Scuola norm. sup. Pisa 2 (IV), 421-468.

Friedman, A. and D. Kinderlehrer (1975): A one phase Stefan problem. Ind. Math. J. 24, 1005-1035.

Fulks, W. B. and R. B. Günther (1969): A free boundary problem and an extension of Muskat's model. Acta Math. 122, 273-300.

Gaipova, A. N. (1968): A homogeneous implicit difference scheme for the solution of an evolutionary equation with phase variations. USSR Comp. Math. math. Phys. 8, (3), 40-53.

Gajewski, H. (1977): Stabilitätsaussagen für einige Aufgaben mit freier Grenze. ZAMM 57, 439-447.

George, J. and P. S. Damle (1975): On the numerical solution of free boundary problems. Int. J. Num. Meth. Engng. 9, 239-245.

Goodman, T. R. and J. J. Shea (1960): The melting of finite slabs. J. Appl. Mech. 27, 16-24.

Hager, W. W. and G. Strang (1975): Free boundaries and finite elements in one dimension. Math. Comp. 29, 1020-1031.

Hansen, E. and P. Hougaard (1974): On a moving boundary problem from biomechanics. J. Inst. Math. Applics. 13, 385-398.

Hill, A. V. (1928): The diffusion of oxygen and lactic acid through tissues. Proc. Royal Soc. London, 104, 39-96.

Hill, C. D. and D. Kotlow (1972): Classical solutions in the large of a two-phase free boundary problem I. Arch. Rat. Mech. Anal. 45, 63-78.

Hoffman, K. -H. (1978): Monotonie bei nichtlinearen Stefan-Problemen. ISNM 39, 162-190.

Höhn, W. (1978): Konvergenzordnung bei einem expliziten Differenzenverfahren zur numerischen Lösung des Stefan-Problems. ISNM 39, 191-213.

Huber, A. (1939): Über das Fortschreiten der Schmelzgrenze in einem linearen Leiter. ZAMM 19, 1-21.

Hunt, C. and N. R. Nassif (1973): On a variational inequality and its approximations in the theory of semiconductory. SIAM J. Numer. Anal. 12, 938-950.

Jerome, J. (1976): Existence and approximation of weak solutions of the Stefan problem with nonmonotone nonlinearities. Lecture Notes Math. 506, 108-156.

Jerome, J. (To appear): Nonlinear equations of evolution and a generalized Stefan problem.

Jiji, L. M. (1970): On the application of perturbation to free-boundary problems in radial systems. J. Franklin Inst. 289, 281-291.

Kamynin, L. I. (1963): On the existence of a solution of Verigin's problem. USSR Comput. Math. math. Phys. 2 , 954-987.

Kinderlehrer, D. and L. Nirenberg (1977). Regularity in free boundary problems. Ann. Scuola norm. sup. Pisa 4 (IV), 373-391.

Kolodner, I. I. (1956): Free boundary problems for the heat equation with applications to problems of change of phase. Comm. Pure Appl. Math. 9 , 1-31.

Kotlov, D. B. (1973): A free boundary problem connected with the optimal stopping problem for diffusion processes. Trans. AMS 184, 457-478.

Kruzhov, S. (1967): On some problems with unknown boundaries for the heat conduction equation. J. Appl. Math. Mech. 31, 1009-1020.

Kyner, W. T. (1959a): An existence and uniqueness theorem for a nonlinear Stefan problem. J. Math. Mech. 8, 483-498.

Kyner, W. T. (1959b): On a free boundary value problem for the heat equations. Quart. J. Appl. Math. 17, 305-310.

Landau, H. A. (1950): Heat conduction in a melting solid. Quart. J. Appl. Math. 8, 81-94.

Lee, Y. F. and B. A. Boley (1973): Melting an infinite solid with a spherical cavity. Int. J. Engng. 11, 1277.

Li-Shang, J. (1965): Existence and differentiability of the solution of a two-phase problem for quasi-linear parabolic equations. Acta Math. Sinica, 15, 6, 749-764.

Lotkin, M. (1960/61): The calculation of heat flow in melting solids. Quart. J. Appl. Math. 18, 79-85.

Magenes, E. (1976): Topics in parabolic equations: some typical free boundary problems. Boundary value problems for linear partial differential equations, Proceedings of the NATO Advanced Study Institute held at Liège, Belgium, Sept. 6-17, 1976; H. G. Garnier ed., D. Reidel Publ. Comp. Dortrecht, 239-312.

Meadley, C. K. (1971): Back diffusion in a finite medium with a moving boundary. Quart. J. Mech. Appl. Math. 24, 43-51.

Meyer, G. H. (1970): On a free interface problem for linear ordinary differential equations and the one-phase Stefan problem. Numer. Math. 16, 248-267.

Mayer, G. H. (1971): A numerical method for two-phase Stefan problems. SIAM J. Numer. Anal. 8, 555-568.

Meyer, G. H. (1973): Multidimensional Stefan problems. SIAM J. Numer. Anal. 10, 522-538.

Meyer, G. H. (1977a): One-dimensional parabolic free boundary problems. SIAM Review 19, 17-34.

Meyer, G. H. (1977b): An alternating direction method for multidimensional parabolic free surface problems. Int. J. Numer. Meth. Engng. 11, 741-752.

Meyer, G. H. (1977c): An application of the method of lines to multidimensional free boundary problems. J. Inst. Math. Applics. 20, 317-329.

Meyer, G. H. (1978): The method of lines for Poisson's equation with nonlinear or free boundary conditions. Numer. Math. 29, 329-344.

Millinazzo, F. and G. W. Bluman (1975): Numerical similarity solutions to Stefan problems. ZAMM 55, 423-429.

Miranker, W. L. (1958): A free boundary value problem for the heat equation. Quart. J. Appl. Math. 16, 121-130.

Miranker, W. L. and J. B. Keller (1960): The Stefan problem for a nonlinear equation. J. Math. Mech. 9, 67-70.

van Moerbeke, R. (1976): On optimal stopping and free boundary problems. Arch. Rat. Mech. Anal. 60, 101-148.

Nigam, S. D. and H. C. Agrawal (1960): A variational principle for convection of heat. J. Math. Mech. 9, 869-884.

Oleinik, O. A. (1960): A method of solution of the general Stefan problem. Sov. Math. Dokl. 1, 1350-1354.

Rasulov, T. M. (1976): Solution of one-dimensional mixed problems for a parabolic system in a region with a variable boundary. J. Diff. Equ. 12, 911-915.

Rogers, J. C. W. (1977): A free boundary problem as diffusion with nonlinear absorption. J. Inst. Math. Applics. 20, 261-282,

Rose, M. (1960): A method for calculating solutions of parabolic equations with a free boundary. Math. Comp. 14, 249-256.

Rubinstein, L. I. (1971): The Stefan problem. Transl. Math. Monographs 27, American Math. Society, Providence, R.I.

Sachs, A. (1975): Zur numerischen Behandlung freier Randwertprobleme parabolischer Differentialgleichungen. ISNM 26, 119-126.

Sackett, G. G. (1971a): Numerical solution of a parabolic free boundary problem arising in statistical decision theory. Math. Comp. 25, 425-434.

Sackett, G. G. (1971b): An implicit free boundary problem for the heat equation. SIAM J. Numer. Anal. 8, 80-95.

Schafferer, D. (1976): A new proof of infinite differentiability of the free boundary in the Stefan problem. J. Diff. Equ. 20, 266-269.

Schatz, A. (1969): Free boundary problems of Stefan type with prescribed flux. J. Math. Anal. Appl. 28, 569-580.

Selig, F. (1956): Bemerkungen zum Stefanschen Problem. Österr. Ing. Archiv 10, 277-280.

Sherman, B. (1965): A free boundary problem for the heat equation with input at a melting interface. Quart. J. Appl. Math. 23, 337-347.

Sherman, B. (1967): A free boundary problem for the heat equation with prescribed flux at both fixed face and melting interface. Quart. J. Appl. Math. 25, 53-63.

Sherman, B. (1970): A general one-phase Stefan problem. Quart. J. Appl. Math. 28, 377-382.

Sherman, B. (1971): General one-phase Stefan problems and free boundary problems for the heat equations with Cauchy data prescribed on the free boundary. SIAM J. Appl. Math. 20, 555-570.

Solomon, A. (1966): Some remarks on the Stefan problem. Math. Comp. 20, 347-360.

Stefan, J. (1889): Über einige Probleme der Theorie der Wärmeleitung. Sitz.-Ber. Wien Akad. Math. Naturw. 98, 473-484.

Tadjbakhsh, I. and W. Liniger (1964): Free boundary problems with regions of growth and decay. Quart. J. Mech. Appl. Math. 17, 141-155.

Tayler, A. B. (1975): The mathematical formulation of Stefan problems. Moving boundary problems in heat flow and diffusion. J. R. Ockendon and W. R. Hodgkins eds., Claredon Press, Oxford, 120-137.

Trenck, W. E. (1959): On an explicit method for the solution of a Stefan problem. J. SIAM 7 , 184-204.

Vasilev, F. P. (1968): The method of straight lines for the solution of a one-phase problem of the Stefan type. USSR Comp. Math. Math. Phys. 8, (1), 81-101.

Ventsel, T. D. (1960): A free boundary problem for the heat equation. Sov. Math. Dokl. 1, 358-361.

Wellford, L. C., Jr., and R. M. Ayer (1977): A finite element free boundary formulation for the problem of multiphase heat conduction. Int. J. Num. Meth. Engng. 11, 933-943.

Wu, T. S. (1966): Bounds in melting slab with several transformation temperatures. Quart. J. Mech. Appl. Math. XIX, 183-195.

Wu, T. S. and B. A. Boley (1966): Bounds in melting problems with arbitrary rates of liquid removal. SIAM J. Appl. Math. 14, 306-323.

Yagoda, H. P. and B. Z. Boley (1970): Starting solutions for melting of a slab under plane or axisymmetric hot spots. Quart. J. Mech. Appl. Math. 23, 225-246.

Institut für Angewandte Mathematik
Albert-Ludwigs-Universität
Hermann-Herder-Str. 10
7800 Freiburg i. Br.
Federal Republic of Germany

The Hodie Method and its Performance for Solving Elliptic Partial Differential Equations

Robert E. Lynch and John R. Rice

1. INTRODUCTION

We consider a new flexible, high accuracy finite difference approximation to the elliptic partial differential equation:

$$Lu = Au_{xx} + Bu_{xy} + Cu_{yy} + Du_x + Eu_y + Fu = f, \quad (x,y) \in R \qquad (1.1)$$
$$u = g, \quad (x,y) \in \partial R$$

where R is a region with piecewise smooth boundary ∂R in the (x,y)-plane and the coefficients A,...,F,f,g are given functions of x and y. A rectangular mesh is put over R and at each mesh point an estimate U for u is obtained as the solution of a finite difference equation. For simplicity of exposition, we assume the mesh is uniform with mesh spacing h; this assumption is not essential to the method though it improves its efficiency in some cases.

To derive the finite difference approximation, consider the origin to be a mesh point; then it plus the eight neighbor mesh points $(\pm h,0)$, $(0,\pm h)$, $(\pm h,\pm h)$ are the nine stencil points of the difference equation. We use U_i, i=0,...,8 to

ISBN 0-12-208360-1

denote the values of the approximation U at these stencil points. We also use a set (x_j,y_j), $j = 1,2,\ldots,J$ of distinct auxiliary (or evaluation) points in the square S_h of side 2h containing the stencil points. With $f_j = f(x_j,y_j)$, the difference equation is

$$L_h U \equiv (1/h^2)\sum_{i=0}^{8} \alpha_i U_i = \sum_{j=1}^{J} \beta_j f_j \equiv I_h f \tag{1.2}$$

and this equation defines the operators L_h and I_h. With an appropriate normalization, $I_h f = f + O(h)$ and thus $I_h f$ is an identity expansion. This leads to the name High Order Differences with Identity Expansion and the acronym Hodie.

After the auxiliary points are chosen, the coefficients α_i, β_j of the difference equation are determined to make the approximation exact on a given linear space of functions (in this paper we use a space of polynomials). Let $s_0,\ldots,s_{J+7}$ denote a basis for the space. The coefficients are chosen to satisfy

$$(1/h^2) \sum_{i=0}^{8} \alpha_i (s_k)_i = \sum_{j=1}^{J} \beta_j (Ls_k)_j, \quad k = 0,\ldots,J+7 \tag{1.3a}$$

together with a normalization equation such as

$$\beta_1 = 1 \tag{1.3b}$$

or some other convenient one. These are the Hodie equations and, except for constant coefficient operators L, the α_i, β_j depend on the location (x_0,y_0) of the central stencil point; we suppress explicit display of this dependence. Note that these equations are local in the sense that only values of quantities defined on S_h enter.

An analysis of this method for ordinary differential equations is given by Lynch and Rice [1979a] and for elliptic partial differential equations by Lynch and Rice [1978] and [1979b]. The purpose of this paper is to discuss the Hodie method's properties, implementation, and performance. For completeness, we summarize some basic theoretical results in Section 2. Section 3 gives a general discussion of the method's computational properties and potential applicability. Section 4 discusses specific implementations which have been made, and the final section summarizes a comparative performance evaluation using the ELLPACK system, Rice [1977].

The Hodie difference equation is similar to the Mehrstellenverfahren ("Hermitian" method), Collatz [1960], if in the Mehrstellenverfahren one replaces derivatives of f with divided differences. The Mehrstellenverfahren determines coefficients by equating coefficients of linear combinations of Taylor's expansions of u and Lu. Young and Dauwalder [1965] give complete details and formulas for one application of the Mehrstellenverfahren for (1.1) and a nine point stencil. Their difference equation is similar to (1.2), but with a linear combination of derivatives of f on the right side. Another difference approximation is given by Rosser [1975] for the Poission equation; he also matches terms in Taylor's series. Rosser replaces derivatives with differences; his $O(h^6)$ method uses values of f at mesh points outside the square S_h.

For ordinary differential equations, the Hodie method gives the same difference equation as Osborn [1967] who generalized the Størmer-Numerov scheme. More recently and independently, Doedel [1976, 1978] presented an essentially equivalent method for ordinary differential equations and he proved a number of convergence results. Both Osborn's and Doedel's approaches lead to less efficient implementations than that of Lynch and Rice [1979a].

2. SUMMARY OF THEORETICAL RESULTS

A complete analysis of the m-th order ordinary differential equation case is given by Lynch and Rice [1979a] which may be summarized as follows:

> The use of J auxiliary points gives truncation and discretization errors of order $O(h^J)$. There are Gauss-type auxiliary points which count as two points.

An example of a Gauss type point occurs in the simplest case where the central stencil point is the single auxiliary point. The usual difference equation for a second order problem can be derived by making the scheme exact on quadratic polynomials and it is automatically exact on cubic polynomials; the discretization error is $O(h^2)$. Any other choice of auxiliary point gives a scheme which is not exact on cubics; the discretization error is then $O(h)$.

The situation for partial differential equations is more delicate and more interesting than for ordinary differential equations. For example, it is apparent from Milne [1953] (see the formula for K in 10. on page 139) that the usual nine point approximation for the Laplacian can be used to obtain $O(h^6)$ accurate approximations to smooth solutions of the Poisson equation by using f, $\nabla^2 f$, $\nabla^4 f$, and $\partial^4 \nabla^2 f/\partial x^2 \partial y^2$ evaluated at the origin; that is, the approximation is exact on the space of polynomials $\mathbb{P}_7$ of degree at most 7. Theorem 11 of Birkhoff and Gulati [1975] establishes that there is no nine point finite difference approximation for the Laplacian which is exact on the space $\mathbb{P}_8$. Thus, there is an upper limit on the order of accuracy of nine point polynomial approximations to partial elliptic partial differential equations in two variables.

The <u>truncation error</u> is defined as the max-norm of $L_h u - I_h Lu$ and the <u>discretization error</u> is the max norm of $u - U$. In the cases we have analyzed, the order of the discretization error is the same as the order of the truncation error. For nine point Hodie approximations exact on polynomial spaces for problems with sufficiently smooth solutions, we have proved:

(a) <u>If $Lu = u_{xx} + u_{yy}$, then the Hodie method gives $O(h^6)$ truncation error</u>.

(b) <u>If $Lu = u_{xx} + c\, u_{yy}$ or $Lu = u_{xx} + 2b\, u_{xy} + u_{yy}$ with constants $b \neq 0$, $c \neq 1$, then the Hodie method gives $O(h^4)$ truncation error</u>.

(c) <u>If $Lu = u_{xx} + 2b\, u_{xy} + c\, u_{yy}$, with constants $b \neq 0$ $c \neq 1$, then the Hodie method gives $O(h^2)$ truncation error</u>.

See Young and Dauwalder [1965] for similar results.

The orders of the truncation errors given in the results above are theoretical upper limits for these simple operators. This follows, as in the theorem of Birkhoff and Gulati [1975], because there are elements in the null space $\mathbb{N}$ of these operators which are polynomials of arbitrarily large degree. If one attempts to construct a Hodie approximation which is exact on $\mathbb{P}_n$, then if n is sufficiently large the coefficients α_i are all equal to zero because the intersection $\mathbb{P}_n \cap \mathbb{N}$ is linearly independent with respect to nine stencil points. This analysis leads to the general result for Hodie approximations which are exact on a linear space S (not necessarily polynomials):

(d) <u>There is no useful Hodie approximation exact on S if $S \cap \mathbb{N}$ is linearly independent with respect to the stencil points</u>.

However, the preceding result might be quite specialized because it is not common that the dimension of $\mathbb{P}_n \cap \mathbb{N}$ for a particular operator L is very large, even when n is arbitrarily large. In particular, we have found a Hodie approximation to

$$Lu = (e^{x+y}u_x)_x + (e^{x+y}u_y)_y + 2(x^2+y^2)e^{x+y}u = 0$$

which has discretization error $O(h^7)$. The exact nature of these higher order Hodie methods is still unclear and it is possible that they might be numerically unstable or lead to badly ill-conditioned systems; however, no such computational difficulties were observed in numerous experiments.

We also have shown:

(e) The difference coefficients α_i, β_j of a high order Hodie approximation differ from those of the standard nine point difference coefficients by $O(h)$.

A similar result is based on the constant coefficient operator L_0 related to the variable coefficient operator L obtained by setting $D = E = F = 0$ and evaluating the functions A, B, C, of L at the central stencil point:

(f) The Hodie coefficients α_i, β_j for L differ from those of L_0 by $O(h)$.

Thus, all the Hodie difference equations, of whatever order of accuracy, are $O(h)$ perturbations of the standard $O(h^2)$ finite difference equations.

We have concentrated on nine point Hodie approximations because they seem to be of the most practical interest. Five point Hodie approximations also exist with similar properties, but the theoretical limit of accuracy is less for these. We have also limited this presentation to a simple case of a second order differential operator. The method and analysis extends to higher order differential operators, to problems with three or more independent variable, to nonuniform meshes, to operators different from elliptic, and to problems with general linear boundary conditions such as

$$Pu + Qu_x + Tu_y = g$$

on curved domains.

In the case of curved domains, the stencil points can be chosen as in Figure 1 and the Hodie difference equation at the boundary is

$$(1/h^2) \sum_{i=0}^{I} \alpha_i U_i = \sum_{j=1}^{J} \beta_j f_j + \sum_{k=1}^{K} \gamma_k g_k. \tag{2.1}$$

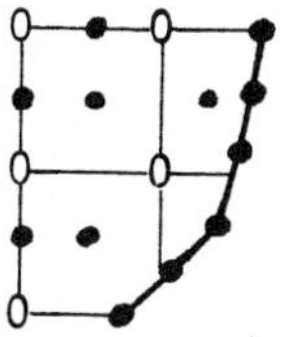

Figure 1. Stencil points 0's, dots indicate interior and boundary auxiliary points

Here I denotes the number of stencil points which are in the interior of the domain and in the square S_h and K is the number of points on the boundary at which the right side, g, of the boundary conditions is evaluated. One might need $I + K = 11$ and so far, we have only been able to obtain $O(h^5)$ truncation error for the Poisson problem in this situation.

Finally, we note that we have shown that all the Hodie approximations discussed above actually exist.

3. EFFICIENCY ESTIMATES OF THE HODIE METHOD

The solution of a partial differential equation by the Hodie method involves two distinct steps. First, one solves the small linear system of Hodie equations (1.3) for the coefficients α, β for each mesh point (except in the case of a constant coefficient operator). Second, one solves the resulting (global) system of difference equations for the values of U. This is a large linear system whose structure is independent of the order of accuracy of the Hodie approximation.

For simplicity of analysis, we assume that the rectangle R is the unit square and the mesh length is $h = 1/(N+1)$ so that there are N^2 interior mesh points. We consider the space S to be the space $\mathbb{P}_m$, of polynomials of degree at most m which has dimension $(m+1)(m+2)/2$. In general, a Hodie scheme which is exact on $\mathbb{P}_m$ is of order $O(h^{m-1})$. The standard finite difference approximation is derived to be exact on $\mathbb{P}_2$, but the method is acutally exact on $\mathbb{P}_3$ because the single auxiliary point (the central stencil point) is a Gauss-type point, hence the method has accuracy $O(h^2)$.

Similarly, it is usually the case that a Hodie scheme exact on $\mathbb{P}_m$ gives $O(h^m)$ accuracy rather than the expected $O(h^{m-1})$; however, below we consider the more general case that $\mathbb{P}_m$ leads to an $O(h^{m-1})$ scheme.

The two steps of the computation can then be stated as: first one solves N^2 local systems to obtain the coefficients α_i, β_j of an $O(h^{m-1})$ difference approximation and, second, one solves one global system of N^2 difference equations. The work of solving the global system is $O(N^4)$ when profile band elimination is used (this work can be reduced when special methods of solution are applicable) and thus, for large enough N, the computation of the Hodie coefficients is negligible since it is $O(N^2)$.

3.1 The Solution of the Hodie Equations.

The rough analysis above suggests that it does not matter much how one solves the local Hodie equations (1.3). But this is not so for two reasons. First, for m greater than 3 (the case of interest) and for small N, say N less than 25, the $O(N^2)$ term for the local systems dominates the $O(N^4)$ term. Consequently, one wants an algorithm which efficiently evaluates the α's and β's. Second, as indicated below, the local system for the α's and β's might be nonstandard because it is either singular or, as h tends to zero, it tends to a singular system.

We now examine, in some detail, the amount of work involved in the evaluation of the difference equation coefficients.

We first note that the system is reducible and that an appropriate choice of basis elements for $\mathbb{P}_m$ separates the problem into two parts. The matrix form of (1.3) is

$$\begin{pmatrix} M_1 & M_2 \\ 0 & M_3 \end{pmatrix} \begin{pmatrix} \alpha \\ \beta \end{pmatrix} = \begin{pmatrix} 0 \\ 0 \end{pmatrix} \tag{3.1}$$

with normalization (1.3b).

For the first nine basis elements, one can use the power basis which spans the set of biquadratic polynomials:

$$s_k(x,y) = (x/h)^p (y/h)^q, \quad p,q = 0,1,2.$$

The factors of $1/h$ make the values of these basis elements either zero or ± 1 at the stencil points. One gets the system $M_1\alpha = -M_2\beta$ to be

$$\sum_{i=0}^{8} \alpha_i (s_k)_i = \sum_{j=1}^{J} \beta_j (Ls_k)_j, \quad k=0,\ldots,8, \tag{3.2}$$

which, <u>once</u> $M_2\beta$ <u>has been determined</u>, can be solved for the α's by using 8 multiplications and 23 additions (henceforth we neglect additions because, typically, one occurs for each multiplication).

For the rest of the basis elements, we choose polynomials which are divisible by either $x^3 - xh^2$ or $y^3 - yh^3$ so that their values are zero at each of the nine stencil points. This gives the homogeneous system $M_3\beta = 0$, which, with the normalization, has J-1 unknowns and is

$$\sum_{j=1}^{J} \beta_j (Ls_k)_j = -(Ls_k)_1, \quad k=9,\ldots,J+8. \tag{3.3}$$

Since $J+8 = (m+1)(m+2)/2$, the cases of interest are

J = 7 with m = 4 to get a 3rd order Hodie method;
J = 13 with m = 5 to get a 4th order Hodie method;
J = 20 with m = 6 to get a 5th order Hodie method;
J = 28 with m = 7 to get a 6th order Hodie method.

Since it is sometimes the case that one obtains one more than the expected order, we do not discuss, below, the case of J = 28.

To determine the β's we use basis elements

$$s_k(x,y) = K_k\, x^p(x^2-h^2)y^q(y^2-h^2)/h^{p+q},$$

then, with $X = x/h$ and $Y = y/h$, one has

$$s_k(x,y) = S_k(X,Y) = K_k\, h^2 X^p(X^2-1)Y^q(Y^2-1). \tag{3.4}$$

Derivatives of s_k with respect to x and y are equal to derivatives of S_k with respect to X and Y when multiplied by an appropriate power of h. Values of S_k and its derivatives are evaluated once and stored and $L(s_k)_j$ in (3.1) (with subscript k suppressed) is:

$$\begin{aligned}(Ls)_j &= A\, S_{XX}(X_j,Y_j) + B\, S_{XY}(X_j,Y_j) + C\, S_{YY}(X_j,Y_j)\\ &\quad +hDS_X(X_j,Y_j) + hES_y(X_j,Y_j) + h^2FS(X_j,Y_j),\end{aligned} \tag{3.5}$$

where A,B,...,F are evaluated at (x_j,y_j).

Typically, we have selected auxiliary points as a subset of points on an equal spaced 9-by-9 grid in the square S_h. The values of X_j and Y_j are then integral multiples of $\pm 1/4$; appropriate choice of the constant K_k makes the values of S_k and its derivatives integers, so no roundoff occurs in these values.

The matrix M_3 in (3.1) is not full, but the precise number of zero elements depends on which, if any, of B,D,E, and F are zero. Some advantage can be taken of the location of the zero entries even in the most general situation, and the work of solving the system is about half that of Gauss elimination for a full matrix. In our operations counts below, we assume that the system for the β's can be solved with $(J-1)^2/6$ multiplications. The resulting multiplication counts for the cases of interest are: 36 multiplication for J = 7, 222 for J = 13, and 819 for J = 20.

The work in evaluating the elements of M_3 after the values of the basis elements and their derivatives and the values of A,B,...,F have been determined is: 216 multiplications for J = 7, 642 for J = 13, and 1452 for J = 20.

The work to form M_2 is similar to that of M_3. There is some savings due to the fact that the simplest basis functions are involved, for example with the basis element $s_0 = 1$, $Ls_0 = F$. The number of multiplications to form $M_2\beta$ once β has been determined is: 122 multiplications for J = 7, 208 for J = 13, and 324 for J = 20.

These results are summarized in Table 1. The entries should be considered only as a general guide since the exact values depend on a specific implementation. However, it is clear that they are substantial numbers compared with the 15 to 20 multiplications required to form the standard $O(h^2)$ difference approximation of Lu = f. For the Hodie method to be advantageous, there must be substantial compensating gains in accuracy. These operations counts also show the potential payoff for clever choices of auxiliary points and basis elements for $\mathbb{P}_m$.

	J = 7 3rd order	J = 13 4th order	J = 20 5th order
Formation of M_3	216	642	1452
Solution of $M_3\beta = 0$	36	222	819
Formation of $M_2\beta$	122	208	324
Solution of $M_1\alpha = M_2\beta$	8	8	8
Total	384	1079	2604

Table 1. Multiplication counts per mesh point for solving the Hodie equations (3.1).

The total multiplication counts for solving the Hodie equations and the difference equations are: $2N^4 + 384N^2$ multiplications for $J = 7$, $2N^4 + 1079N^2$ for $J = 13$, and $2N^4 + 2604N^2$ for $J = 20$. When $J = 13$, the work $2N^4$ of solving the global system is equal to the work $1079N^2$ for solving the set of local systems for N about 24. According to this, one would expect the error to be decreasing as N^{-4}, for this fourth order method, as the work (execution time) is increasing at the rate of N^2 for N up to about, say, 20 (400 interior mesh points). For N larger than about 24, the work increases as N^4. One expects that the graph of log(error) versus log(execution time) has slope -2 for small values of N and slope -1 for large values of N. For the ordinary $O(h^2)$ scheme, the error is decreasing as N^{-2}, so one expects that the slope is -1/2.

In all of this analysis, we have ignored the work of evaluating the coefficient functions A,B,...,F and the right side f. This $O(N^2)$ work can be substantial for $J = 20$ or for auxiliary points distributed in a general way; obvious efficiencies can be gained by picking the auxiliary points appropriately so that, for example, some of the function values can be used in more than just one of the systems (3.1).

A similar, but less precise, analysis has been made for 3-dimensional problems on the unit cube. The interesting choices of J are: 9 (3rd order Hodie method), 30 (4th order), and 58 (5th order). The corresponding multiplication counts are given in Table 2. The values in Table 2 are coefficients of N^3 because a local system must be solved for this many interior mesh points. The multiplication count for solving the large system of difference equations is $2N^7$ when band elimination is used and one sees that the relative effect of the local computations has changed sharply compared with the 2-dimensional case. Note that the work $2N^7$ in solving the global system by a direct method is so large that the use of special methods should be considered.

	J = 9 3rd order	J = 30 4th order	J = 58 5th order
Formation of $M_3\beta$	490	7,400	29,000
Solution of $M_3\beta = 0$	40	3,700	32,000
Formation of $M_2\beta$	700	3,000	6,000
Solution of $M_1\alpha = M_2\beta$	28	28	28
Total	1,258	14,128	67,028

Table 2. Estimated multiplication counts per mesh point for solving (3.1) for a 3-dimensional problem. Number of additions is about the same except for the fourth row: 135 additions are required with the 28 multiplications.

In Tables 3 to 6, we list the multiplication counts and the storage required for the total solution process for four types of problems. Counts are given for the Hodie method and, for comparison, the collocation method with bicubic Hermite polynomials. This collocation method is a 4th order method.

VARIABLE COEFFICIENTS, TWO DIMENSIONAL PROBLEMS				
	Hodie			Collocation
N	3rd order $2N^4+384N^2$	4th order $2N^4+1079N^2$	5th order $2N^4+2604N^2$	4th order $64(N+1)^4+320(N+1)^2$
5	11/.25	28/.25	53/.25	88/7
10	58/2	127/2	226/2	950/40
20	473/16	751/16	1,145/16	12,500/295
50	13,460/250	15,198/250	17,660/250	433,000/4,000

Table 3. Estimates of multiplications and storage counts for two dimensional, variable coefficient problems. The format is (thousands of multiplications)/(thousands of words of storage) where the slash, /, is a separator. The work of evaluation of the coefficients A,B,...,F,f is not included.

CONSTANT COEFFICIENTS, TWO DIMENSIONAL PROBLEMS				
	Hodie			Collocation
N	3rd order $2N^4+7N^2$	4th order $2N^4+13N^2$	5th order $2N^4+20N^2$	4th order $64(N+1)^4$
5	1.4	1.5	1.7	83
10	21	21	22	925
20	324	324	324	10,500
50	12,500	12,500	12,500	400,000

Table 4. Estimates of multiplication counts for two dimensional problems with A,B,...,F constant, f variable. The storage is the same as in Table 3.

VARIABLE COEFFICIENTS, THREE DIMENSIONAL PROBLEMS				
	Hodie			Collocation
N	3rd order $2N^7+1258N^3$	4th order $2N^7+14128N^3$	5th order $2N^7+67028N^3$	4th order $1024(N+1)^4+4100(N+1)^3$
5	314/6	1,922 /6	8,534/6	287,000/995
10	20,126/200	34,128/200	87,028/200	2^{+7}/20,600
20	2.6^{+6}/6400	2.7^{+6}/6,400	3.1^{+6}/6,400	2^{+9}/523,000

Table 5. Estimates of multiplication and storage counts for three dimensional, variable coefficient problems. The format is the same as in Table 3. 2.6^{+6} denotes 2.6×10^6.

CONSTANT COEFFICIENTS, THREE DIMENSIONAL PROBLEMS				
	Hodie			Collocation
N	3rd order $2N^7+9N^3$	4th order $2N^7+30N^3$	5th order $2N^7+58N^3$	4th order $1024(N+1)^7$
5	158	160	164	287,000
10	20,009	20,030	20,058	2^{+7}
20	2.6^{+6}	2.6^{+6}	2.6^{+6}	2^{+9}

Table 6. Estimates of multiplication counts for three dimensional problems with A,B,...,F constants and f variable. The storage is the same as in Table 5.

3.2 Nonuniform Spacing and Curved Boundaries.

One feature of the Hodie method is that only a relatively small perturbation is required at mesh points near curved boundaries. One is already set up to compute the difference approximations at each mesh point and similar Hodie equations (2.1) must be solved next to the boundary. Of course, the equations are different so a different subprogram would be called to solve them, but this would not greatly change the structure of the algorithm nor the amount of computation.

Curved boundaries prevent one from using precomputed values of the basis functions and their derivatives. This increases the work of forming the equations (3.1) and solving them by about a factor of 3. Thus, the multiplication count for a fourth order Hodie method at such a mesh point is about 3300 instead of 1079. This seems to be a modest price to pay for being able to have high accuracy. The classical approaches to obtaining high order difference approximations are clumsy and tedious even for second order accruacy and practically hopeless for 4th order. In general, nonuniform spacing increases the work in the same way, but we have found an automatic way of locally changing coordinates to obtain a uniform mesh at the expense of multiplying the coefficient functions A,...,F by functions which depend on the nonuniform spacing.

This rough analysis of Hodie methods makes them look very promising because their work (cost or execution time) increases asymptotically as $2N^4$ independent of their order of accuracy. But there are four caveats:

(a) Multiplication counts (especially rough ones like these) are not completely reliable guides as to how efficient a method is actually going to be.

(b) We have assumed there is no difficulty in obtaining, a priori, suitable auxiliary points.

(c) Some phases of the calculations might be ill-conditioned.

(d) We might need a very small h before the asymptotic theoretical analysis outlines in Section 2 becomes applicable.

Thus, it is essential that actual implementations of the Hodie method be examined and experimental tests made before its merit can be judged reliably.

4. IMPLEMENTATIONS OF THE HODIE METHOD

An implementation (1976) of the Hodie method for ordinary differential equations, together with experimental results, is reported on by Lynch and Rice [1979a]. One may summarize that implementation by stating that everything worked out as one would hope and no unexpected complications arose, even for 10th, 12th, and 14th order methods. Even though the Hodie method appears to be one of the most efficient for 2-point boundary value problems, it does not have a marked advantage over other methods for ordinary differential equations, in contrast to the case of partial differential equations.

4.1 Exploratory Implementations for Second Order Elliptic Problems.

Shortly after the method was formulated in 1975 for partial differential equations, an exploratory program was written to test the method. Its objectives were to measure the observed rates of convergence and to check the general feasibility of the method; no attempt at computational efficiency was made. An important discovery made was that the usual algorithms for linear systems are not adequate for solving the Hodie equations (1.3) for the case of partial differential equations (there is no difficulty for the case of ordinary differential equations). For simple operators, this system is sometimes singular, but consistent, i.e., row elimination leads to trivial equations: 0 = 0. Library linear equations solvers fail for such systems and we found generalized inverse routines not only very much slower, but also not completely reliable.

This implementation evolved into a "general" Hodie program with which the order and number of polynomial basis elements could be raised and the choice of auxiliary points made in various ways. Many problems were solved with this program, enough to give confidence that the Hodie method was feasible and no additional unexpected difficulties arose.

The main purpose of the experiments was to verify that the behavior of the discretization error as a function of N was as predicted and to explore the nature of the limit on the order of accuracy of the Hodie method. But the question about the upper limit is still not answered. However, there seems to be little doubt about the feasibility of the method for 3rd and 4th order accuracy in general and for 6th order for the case that $A = C$, $B = 0$, and we have found some examples of 7th order accuracy.

4.2 FFT-9, a Fast Fourier Transform Hodie Method.

One can obtain, explicitly, the coefficients for a 4th order Hodie approximation to the coefficient problem

$$A\,u_{xx} + C\,u_{yy} + F\,u = f, \quad A,C,F \text{ constants},$$

on a rectangle with uniform mesh and similarly a 6th order method for the Poisson problem $u_{xx} + u_{yy} = f$, see Houstis and Papatheodorou [1977]. The Fast Fourier Transform can be applied to the resulting difference equations as is done by Houstis and Papatheodorou [1979] in the algorithm FFT9, which is available in ELLPACK. The comparative experiments of Houstis and Papatheodorou [1977] and [1979] show that this is the most efficient "Fast Poisson Solver" currently available.

4.3 HOLR 27-POINT, a Three Dimensional $O(h^6)$ Poisson Solver.

An explicit 27-point $O(h^6)$ Hodie method for the Poisson problem $u_{xx} + u_{yy} + u_{zz} = f$ on a cube with uniform mesh has been obtained by Lynch [1977a, 1977b]. This algorithm was implemented by Ron Boisvert and Lynch in the ELLPACK system as HOLR 27-POINT. This algorithm performs well and as expected; there are no other algorithms in ELLPACK (or generally available) with which to compare its performance.

The difference approximation used in HOLR 27-POINT is $L_h U = I_h f$ where the operators L_h and I_h are

$$L_h U = [-128U + 14 \sum_{r^2=h^2} U + 3 \sum_{r^2=2h^2} U + \sum_{r^2=3h^2} U]/(30h^2)$$

where the summations are over values of U at mesh points and

$$I_h f = [280f + 8 \sum_{r^2=h^2} f + 48 \sum_{r^2=3h^2/4} f + \sum_{r^2=3h^2} f]/720$$

where the summations are over values of f at auxiliary points which consist of some mesh points and the 8 points $(\pm h/2, \pm h/2, \pm h/2)$.

The difference equations are solved by the tensor product method in $O(N^4)$ operations (in contrast, band elimination requires $2N^7$ operations) and f is evaluated about twice per mesh point, i.e., $2N^3$ times. Since the error is decreasing as N^{-6}, the error is $O([\text{cost}]^{-3/2})$.

4.4 HOLR 9-POINT, a Two Dimensional Generalized Helmholtz Solver.

This implementation by Ron Boisvert treats the equation

$$A(x,y)\, u_{xx} + C(x,y)\, u_{yy} + F(x,y)\, u = f(x,y)$$

on a rectangle. It is part of ELLPACK as HOLR 9-POINT; the keywords CONSTANT COEFFICIENTS directs the program to take advantage of the reduction in work when A, C, and F are constants. Both 4th and 6th order options are available. The $O(h^4)$ constant coefficient options treats the same problem as FFT9 and the principal difference is that HOLR 9-POINT uses Gauss elimination instead of the Fast Fourier Transform.

HOLR 9-POINT has three options for the auxiliary points for the $O(h^4)$ constant coefficient case. They are indicated in Figure 2. In these cases, the special nature of the problem has been taken advantage of so that only 5 auxiliary points are needed, rather than the usual 13 required for variable coefficients.

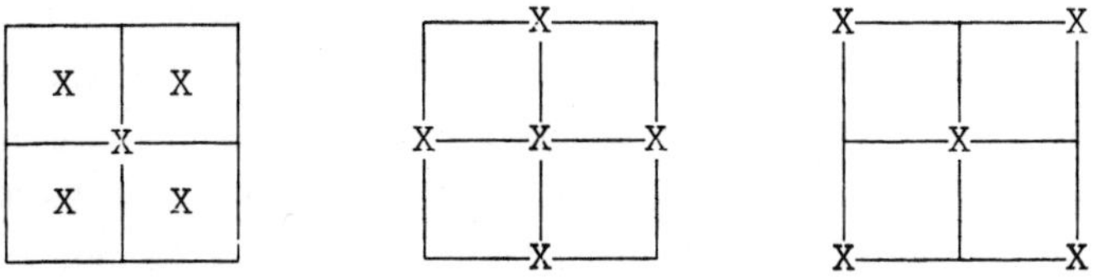

(0) (1) (2)

Figure 2. The three choices of auxiliary points available in HOLR 9-POINT for the $O(h^4)$ constant coefficient case.

The choice (0) is the default for HOLR 9-POINT while FFT9 uses choice (1). The choice (0) requires two evaluations of f per mesh point and the other two require one. But (0) is sufficiently more accurate that it is the most efficient of the three in normal situations.

4.5 The Choice of Auxiliary Points.

It has become apparent that one of the key points with the Hodie method is how to make a good choice of the auxiliary points. This choice significantly affects both efficiency and the accuracy of the resulting method. For example, choice (2) of Figure 2 gives an error which is rather consistantly an order of magnitude worse than choice (0), even though both give $O(h^4)$ accuracy. Choice (2) gains in efficiency over choice (0) because it evaluates f half as many times, but this does not compensate for its larger errors, see Boisvert [1978].

If the auxiliary points are chosen at random, then with probability 1, they give a useful Hodie method. One can make the analogy with polynomial interpolation in the plane. If one chooses 3 points at random, then the probability is 1 that they are not colinear and thus interpolation by a linear polynomial, $p(x,y) = a + bx + cy$, is possible. The analogy carries somewhat further as follows. Once the polynomial degree becomes modest (3 or 4), there is no simple test to apply to a point set in the plane to see if the polynomial interpolation problem is solvable. Likewise, there is no simple test to apply to a point set to see if it defines a useful set of auxiliary points for a Hodie method. Furthermore, in both cases, point sets that are close to the exceptional point sets give ill-conditioned problems and must be avoided. The variable coefficient part of HOLR 9-POINT takes care to avoid ill-conditioned as well as singular choices of auxiliary points and this involves substantial computations not considered in the multiplication counts of the previous section.

4.6 <u>Techniques for Solving the Hodie Equations</u>.

The difficulty in solving the Hodie equations to obtain the α's and β's and the significance of the theoretical results outlined in Section 2 can be seen by reference to equation (3.5). As h tends to zero, the contributions due to D,E, and F tend to zero and the right side of (3.5) tends to

$$A(0,0)\ S_{XX}(X_j,Y_j) + B(0,0)\ S_{XY}(X_j,Y_j) + C(0,0)\ S_{YY}(X_j,Y_j).$$

Thus, the system (3.3) tends to a singular system. We give a single example. For L the Laplacian, the system (3.3) can be solved with J = 27 to give an approximation exact on $\mathbb{P}_7$. The system of 27 equations in (3.3) has rank 20, but this singular system is consistent. Thus, to obtain accurate values of the Hodie coefficients for a general L, one must use specific information about the system in order to construct a reliable algorithm which works in all cases.

Recently, Lynch has completed a detailed analysis of the computation of the Hodie coefficients for the variable coefficient equation

$$A\ u_{xx} + C\ u_{yy} + D\ u_x + E\ u_y + F\ u = f$$

for the case of a uniform mesh. A reliable algorithm results if one introduces the symmetric parts of the β's, the basis elements, and the coefficients of the differential operator. For example, instead of solving for the β's, one solves for

$$b_2 = [\beta(h,0)+\beta(-h,0)]/2, \quad b_3 = [\beta(h,0)-\beta(-h,0)]/2$$

$$b_6 = [\beta(h/2,h/2)+\beta(-h/2,h/2)+\beta(-h/2,-h/2)+\beta(h/2,-h/2)]/4,$$

and so on; one can use the b's throughout and one does not have to compute the β's. The use of the symmetric parts reduces, somewhat, the computation and, most important, leads to a predetermined pivot strategy. For example, for an $0(h^4)$ scheme, pivots are preassigned for 9 of the 12 equations. The last 3 equations are either row reduced to 0 = 0 or to a nonsingular system and nonzero pivots are easily found.

Although this algorithm has not yet been incorporated into ELLPACK, it has been used with ELLPACK and numerous tests indicate that it works as expected.

5. THE PERFORMANCE OF THE HODIE METHOD

The ELLPACK system is a research tool for the evaluation of numerical methods for elliptic partial differential equations. We summarize here the results of a comparison of FFT9, HOLR 9-POINT, P3C1 COLLOCATION, and 5-POINT STAR. The P3C1 COLLOCATION uses bicubic Hermite polynomials and is an $O(h^4)$ method; 5-POINT STAR is a standard five point $O(h^2)$ finite difference algorithm. In addition, we give some results for HOLR 27-POINT and the algorithm mentioned in Section 4.6. Except for FFT9 and HOLR 27-POINT, profile band elimination was used to solve the system of difference equations.

An evaluation in a wider context is given by Houstis, Lynch, Papatheodorou, and Rice [1978]. In this, results are shown for 5-POINT STAR, P3C1 COLLOCATION, and the Galerkin and the Least Squares methods, the latter three use bicubic Hermite polynomials. In this comparison, P3C1 COLLOCATION was judged to be the best of the four methods; in particular, it is substantially more efficient than Galerkin or Least Squares and gives comparable accuracy.

A population of second order elliptic equations on rectangles has been constructed by Houstis and Rice [1978a] with over 40 different problems, many of which have 1 to 3 parameters. This is a substantial problem set which represents both "real world" and "mathematical" behaviors and which contains examples of most phenomena that are usually considered. The experimental procedures are discussed in more detail in Houstis and Rice [1978b].

For the results we report on here, a subset of 15 problems was chosen where HOLR 9-POINT is applicable. Parameters were chosen for some of these problems to produce a total of 30 partial differential equation problems. The characteristic features of these 30 problems are shown in Figure 3. See Houstis and Rice [1978a] for more precise definitions of these features. One sees from Figure 3 that this set of 30 problems represents problems of:

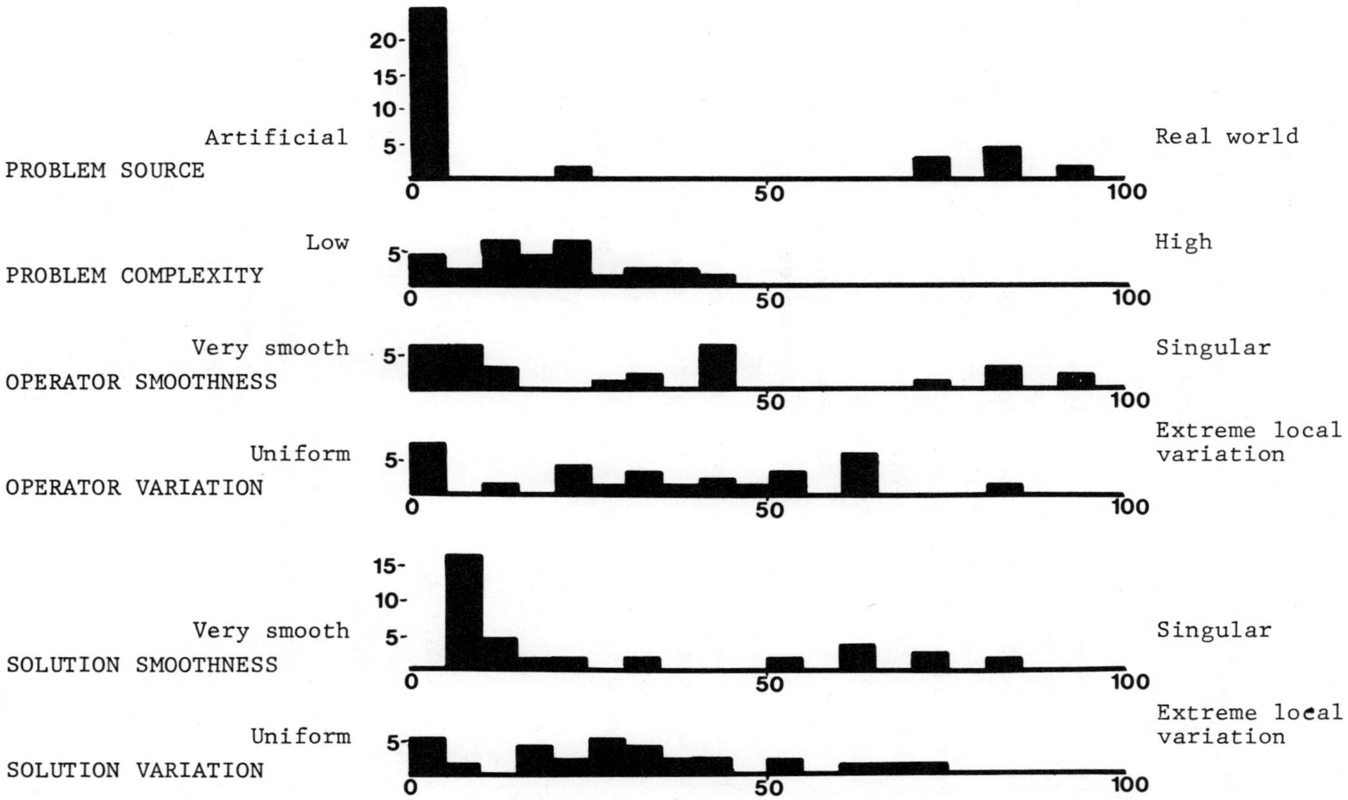

Figure 2. Features of the set of 30 problems used in the performance evaluation (see Houstis and Rice [1978a]). Heights of bars indicate number of problems from the set.

(a) low to moderate complexity,

(b) mostly smooth and uniform operators but with some singular or substantial local variation behaviors present,

(c) a wide range of solution behavior from very smooth to strongly singular, from very uniform to sharp local variations.

In addition to the features shown in Figure 3, we note that

(d) all the boundary conditions are Dirichlet,

(e) eight problems have variable coefficients,

(f) FFT9 is not applicable to 4 problems because of a technical restriction of the domain to [0,a]x[0,b].

Fairly extensive data have been gathered from this comparison effort and are available from the authors. Here we present a summary of the results as:

(a) selected graphs of maximum error versus execution time, this shows the performance of the methods as measured by cost;

(b) selected graphs of maximum error versus storage used;

(c) a table which gives the performance ranks for all 30 problems.

Subsets of the following values of N were used to obtain the points on the graphs, the mesh spacing h is proportional to 1/(N+1)

FFT9:	N = 7,15,31;
HOLR 9-POINT:	N = 1, 3, 5, 7, 9,11,13,15,17,19,21;
P3C1 COLLOCATION:	N = 1, 3, 5, 6;
5-POINT STAR:	N = 3, 7,11,15,19,23;
HOLR 27-POINT:	N = 1, 3, 5, 7, 9.

Only Figure 5a contains all of these points for the first four methods. For HOLR 9-POINT and HOLR 27-POINT, the case N = 1 gives only one interior mesh point.

The computation was done on Purdue University's CDC 6500 Computer with single precision arithmetic which uses floating point numbers with about 15 decimal digits.

The vertical logarithmic axes of the graphs gives the maximum error. The horizontal logarithmic time axes give execution time in seconds. This time includes the time for discretization, indexing, and solution of the linear system;

times for preprocessing and output are not included. In ELLPACK, the discretization modules form a packed array which contains the coefficients and right sides of the difference equations, the indexing modules transform this array to a band matrix. In the graphs which give the memory required, the horizontal linear memory axis includes storage for the algorithm modules and associated variables, but it does not include the space for other parts of the ELLPACK system.

Below we list the problems used to obtain the results displayed in Figures 4,5, and 6. The Problem Number and the Parameters refer to those of the population of Houstis and Rice [1978a].

Figures 4a and 4c. Problem 5 with parameter $\alpha = 64$:

$$4u_{xx} + u_{yy} - 64u = f, \quad u = 2(x^2-x)[\cos(2\pi y) - 1].$$

The curves illustrate behavior typical of smooth, well behaved problems with constant coefficients. In Figure 4a, 5-POINT STAR performs comparatively better than in other cases of this class of problems. As expected, FFT9 gives the best performance. Note that the memory requirements for 5-POINT STAR and P3C1 COLLOCATION are significantly greater than for HOLR 9-POINT and FFT9.

Figures 4b and 4d. Problem 7 with parameter $\alpha = 20$:

$$u_{xx} + u_{yy} - 64u = f,$$
$$u = \cosh(10x)/\cosh(10) + \cosh(20y)/\cosh(20).$$

The curves illustrate behavior typical of boundary layer problems. The relative ranks of the methods for both execution time and memory are the same as for smooth, well behaved problems.

Figure 5a. Problem 17 with parameters $\alpha = 8$, $\beta = 5$:

$$u_{xx} + u_{yy} = f$$
$$u = \sin(x-y+.5) + \exp\{-y^2-[8(5x)^3/(1+5^3x^3)]^2\}.$$

The solution has a sharp wave front: there is a 50% drop in value in an interval of width 4/100. Coarse meshes give errors of 100 to 300 percent. Only FFT9 achieves an error smaller than 10% for this difficult problem.

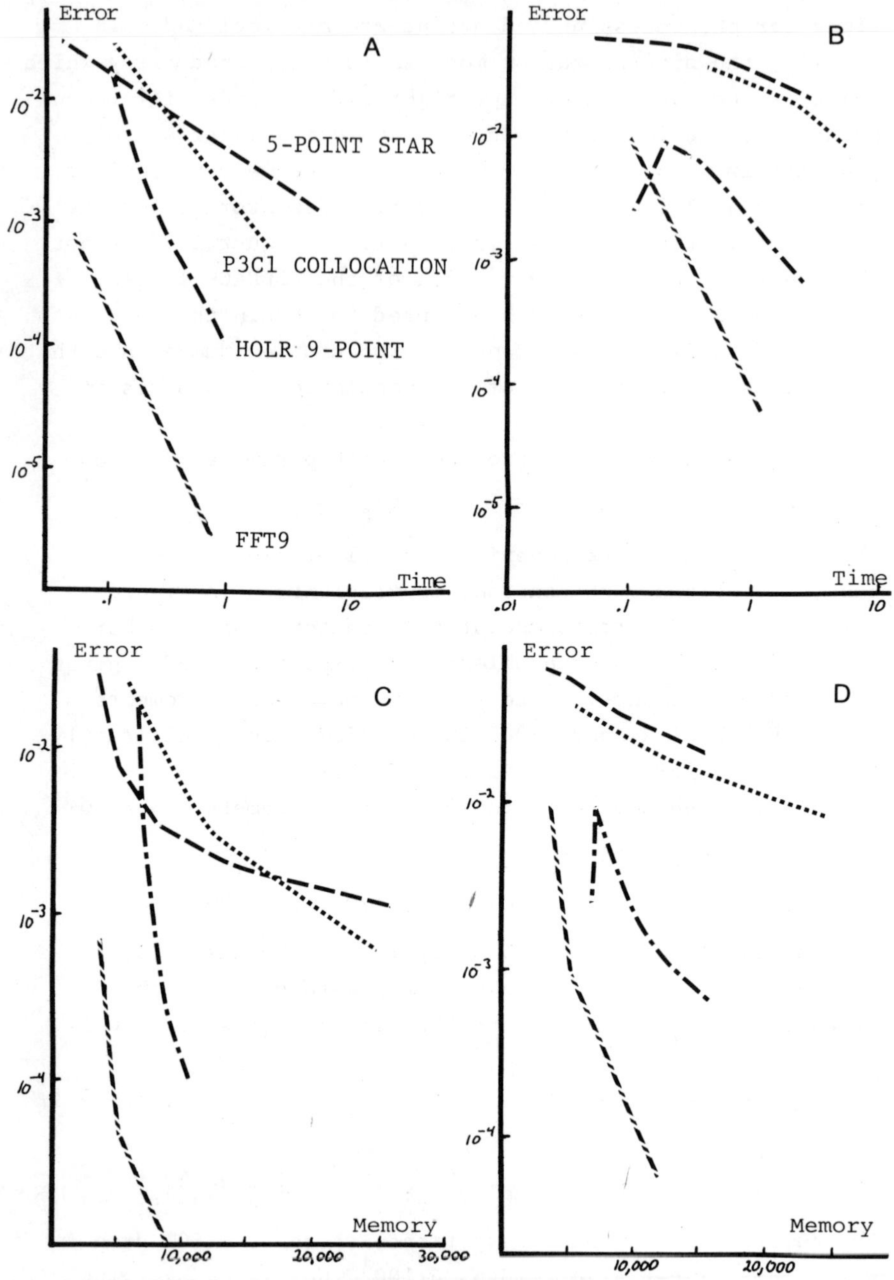

Figure 4. Performance of four methods with respect to execution time and memory. The curves are for the methods as indicated in Figure 4a. Problems: 5 (a & c), 7 (b & d).

<u>Figure 5b</u>. Problem 9 with parameters $\alpha = 50$, $\beta = .5$:

$$u_{xx} + u_{yy} = f, \quad u = \exp\{-50[(x-.5)^2 + (y-.5)^2]\}.$$

The solution has a sharp peak at $x = y = .5$ where its value is unity; it is essentially zero at distances greater than .25 from this point. The performance of P3C1 COLLOCATION is more adversely affected than the other methods. The relative rankings of the other three methods is the same as for smooth, well behaved problems.

<u>Figure 5c</u>. Problem 38 with parameter $\alpha = 3$:

$$u_{xx} + u_{yy} = f, \quad u = (xy)^{3/2}.$$

This illustrates the effect of a singularity in the second derivative of the solution. The relative ranks are the same as for smooth, well behaved problems, but the rates of convergence are less. P3C1 COLLOCATION outperforms HOLR 9-POINT because this problem has homogeneous boundary conditions, a feature which significantly improves the efficiency of P3C1 COLLOCATION. FFT9 did not work on this problem because it requires the value of the right side f at point on the boundary where f is infinite; however, in a similar problem in which the infinite value was set equal to zero, FFT9 substantially outperforms the other 3 methods.

<u>Figure 5d</u>. Problem 39 with parameters $\alpha=3$, $\beta=6$, $\gamma=2$:

$$Au_{xx} + Cu_{yy} + Fu = f, \quad A = 2 + (y-1)\exp(-3y^4),$$

$$C = 1 + 1/[1+(2x)^6], \quad F = 2[x^2-x + (y-.3)(y-.7)],$$

$$u = (x+y^2)/[1+(2x)^5]+(y-1)(1+x)\exp(-3y^4)+2(x+y)\cos(xy).$$

This shows the effect of a variable coefficient, moderately complex problem. FFT9 does not work for this problem because of its variable coefficients. Four other versions of Problem 39, with different parameter values, were also solved and for these, the behavior of HOLR 9-POINT was more erratic than shown here. This figure also shows results for the recent $O(h^4)$ implementation of Lynch mentioned in Section 4.6; it runs about twice as fast as HOLR 9-POINT and it outperforms the other methods for N greater than 5. In addition, this implementation performs

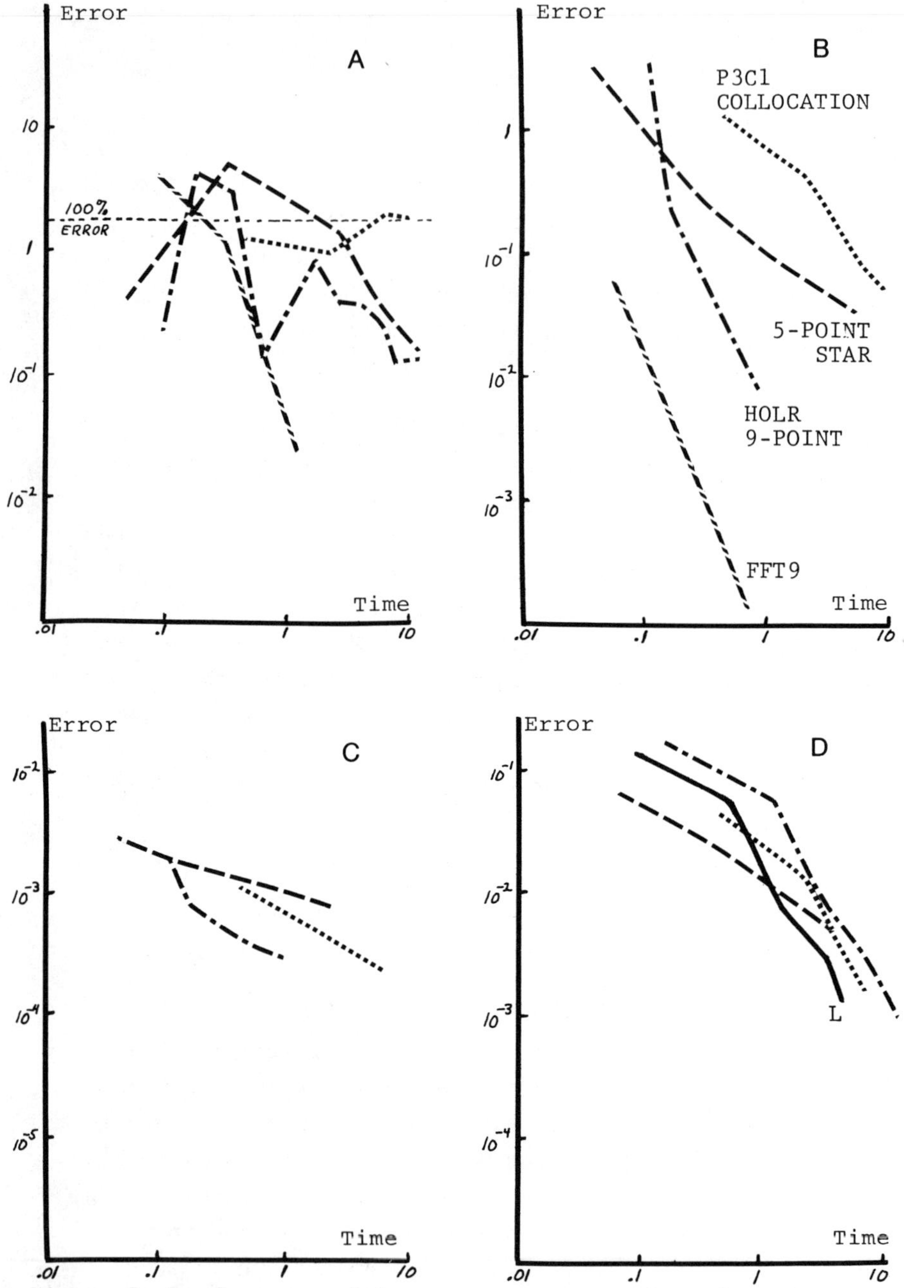

Figure 5. Performance with respect to execution time. Curves for methods as indicated in Figure 5b. Figure 5d shows results for method of Section 4.6, indicated by L. Problems: 17 (a), 9 (b), 38 (c), and 39 (d).

as expected from the operation counts in Section 3 when one neglects the time required to evaluate the functions A,C,F, and f. Specifically, extrapolation from the times for the discretization and solution of the global system predicts that these times are equal for N = 25. However, the time required to evaluate the functions for this problem require about twice the time which is needed to set up and solve all of the Hodie equations to get the coefficients of the difference equations. Consequently for this problem when the time for the function evaluations is included, the discretization time is predicted to be the same as the solution time for N = 39.

<u>Figure 6</u>. $u_{xx} + u_{yy} + u_{zz} = f$,

$u = (x^2-x)(y^2-y)(z^2-z)\exp(x+y+z)$

This illustrates the performance of HOLR 27-POINT for a very simple problem. As expected for this $O(h^6)$ method, the error is reduced by a factor of 1000 for an increase of 100 in execution time.

<u>Table 7</u>. Table 7 simply lists the computational efficiency rankings of the methods as perceived by looking at graphs similar to Figures 4 and 5. It is clear that FFT9, if it is applicable, is the method to choose independent of any problem features present in this set. Furthermore, HOLR 9-POINT consistently ranks second and P3C1 COLLOCATION ranks ahead of 5-POINT STAR, but not by so great a margin.

For the 8 variable coefficient problems where FFT9 is not applicable, the data are inadequate because 5 of the 8 are versions of Problem 39. HOLR 9-POINT does not perform as well as hoped. The other two methods are fairly close and, in fact, P3C1 COLLOCATION would consistently outperform 5-POINT STAR if all of these problems were brought into homogeneous boundary condition form. For these 8 problems 5-POINT STAR performs best overall. Table 7 does not include the rank of Lynch's new algorithm mentioned in Section 4.6 because it has not yet been used on all the problems. Figure 5d shows that it is better than 5-POINT STAR for this version of Problem 39 and, as indicated above,

PROBLEM	PARAMETERS α β γ	FFT9	9-PT HOLR	P3C1	5-PT STAR	REMARKS
3		1	2	3	4	Smooth, well behaved.
4	3	1	3	2	4	$(xy)^{3/2}$ in solution.
4	5	1	1	3	4	$(xy)^{5/2}$ in solution.
5	5	1	2	3	4	Smooth, well behaved.
5	8	1	2	3	4	Smooth, well behaved.
5	10	1	2	3	4	Smooth, well behaved.
5	64	1	2	3	4	Well behaved.
6		-	3	1	2	Smooth but complicated solution.
7	20	1	2	3	4	Boundary layer, P3C1 and 5-PT are close.
8		1	2	3	4	Square wave front, P3C1 and 5-PT cross.
9	50, .5	1	2	4	3	Pronounced peak in solution.
9	100, .5	1	2	4	3	Very pronounced peak in solution.
11	2π	1	2	4	3	6 oscillations, P3C1 and 5-PT cross.
11	5π	1	2	4	3	13 oscillations, coarse meshes give BIG errors.
17	1, 2	1	2	4	3	Rather well behaved.
17	5, 3	1	2	4	3	Pronounced wave front ridge in solution.
17	8, 5	1	2	4	3	Sharp ridge and valley in solution.
20		-	2	3	1	Sharp peak, variable coefficients.
20	0	-	3	2	1	Sharp peak, variable coefficients.
33		*	1	2	3	$\nabla u = -1$ (nearly)

34		*	1	3	2	Smooth, well behaved.
35		*	1	2	3	Smooth, well behaved.
38	3	*	1	2	3	$(xy)^{3/2}$ in solution.
38	5	1	1	3	3	$(xy)^{5/2}$ in solution, P3C1 & 5-PT cross.
38	7	1	2	4	3	$(xy)^{7/2}$ in solution, P3C1 & 5-PT cross.
39	.5, 3,10	-	3	2	1	Well behaved, variable coefficients, HOLR not right.
39	1, 2,.5	-	3	1	2	Well behaved, variable coefficients, HOLR not right.
39	3, 6, 2	-	3	2	1	Some complexity, HOLR and P3C1 cross 5-PT, variable coefficients.
39	10,11, 0	-	3	2	1	Some complexity, variable coefficients.
39	23, 2, 1	-	3	2	1	Fairly well behaved, variable coefficients.

Table 7. Efficiency ranks of FFT9, HOLR 9-POINT, P3C1 COLLOCATION, and 5-POINT STAR (see Houstis and Rice [1978a]). FFT9 was not applied in some cases because of a technical difficulty, these cases are indicated by * .

it behaves as predicted by the operation counts of Section 3.

In assessing these results, one should keep in mind that there is a great difference in the generality of these methods. P3C1 COLLOCATION takes no advantage of any features of the problem except homogeneous boundary conditions. For a general operator and a fixed N, its execution time is the same for a uniform or nonuniform mesh; the use of a nonuniform mesh can reduce the error for some problems, such as the boundary layer problem of Figure 4b and problems with sharp peaks and wave fronts as in Figures 5a and 5b. HOLR 9-POINT would be somewhat affected by the use of a nonuniform mesh and 5-POINT STAR would be greatly affected because its order would be reduced from $O(h^2)$ to $O(h)$. FFT9, of course, is not general enough to cover all of these problems.

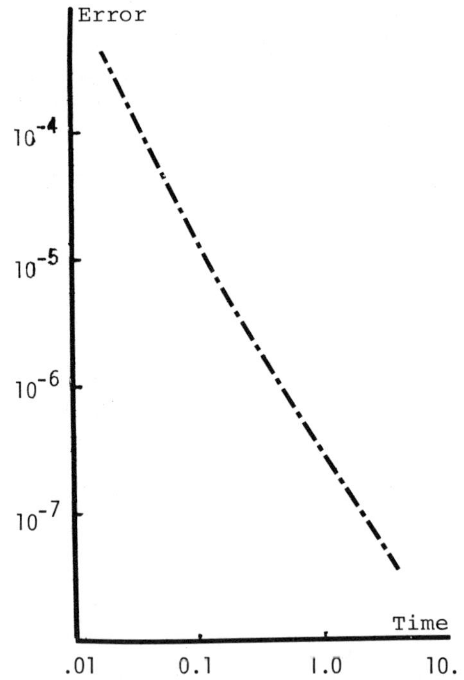

Figure 6. Performance of HOLR 27-POINT with respect to execution time for problem given in the text.

REFERENCES

1. G. Birkhoff and S. Gulati (1975), Optimal few-point discretizations of linear source problems, SIAM J. Numer. Anal. 11 700-728.
2. R.F. Boisvert (1978), The effect on accuarcy of the placement of auxiliary points in the Hodie method for the Helmholtz problem, Dept. Computer Science Report CSD-TR 266, June 1978, Purdue Univ.
3. L. Collatz (1960), Numerical Treatment of Differential Equations, 3rd Ed. Springer Verlag.
4. E.J. Doedel (1976), The construction of finite difference approximations to ordinary differential equations, Appl. Math. Report, Calif. Inst. Tech.
5. ________ (1978), The construction of finite difference approximations to ordinary differential equations, SIAM J. Numer. Anal. 15 450-465.
6. E.N. Houstis, R.E. Lynch, T.S. Papatheodorou, and J.R. Rice (1978), Evaluation of numerical methods for elliptic partial differential equations, J. Comp. Physics 27 (to appear).
7. E.N. Houstis and T.S. Papatheodorou (1977), Comparison of fast direct methods for elliptic problems, in Advances in Computer Methods for Parital Differential Equations II (R. Vishnevetsky, Edit.) IMACS, New Brunswick, N.J., 46-52.
8. ________ (1979), Alogrithm 5xx, FFT9: Fast solution of Helmholtz type partial differential equations, ACM Trans. Math. Software (to appear).
9. E.N. Houstis and J.R. Rice (1977), Software for linear elliptic problems on general two dimensional domains, in Advances in Computer Methods for Partial Differential Equations II (R. Vishnevetsky, Edit.), IMACS, New Brunswick, N.J., 7-12.
10. ________ (1978a), A population of partial differential equations for evaluating methods, Dept. Computer Science Report CDS-TR 263,

May 15, 1978, Purdue Univ.

11. E.N. Houstis and J.R. Rice (1978b), An experimental design for the computational evaluation of elliptic partial differential equations, Dept. of Computer Science Report CSD-TR 264, May, 1978, Purdue Univ.
12. R.E. Lynch (1977a), $0(h^6)$ accurate finite difference approximation to solutions of the Poisson equation in three variables, Dept. of Computer Science Report CSD-TR 221, February 15, 1977, Purdue Univ.
13. ________ (1977b), $0(h^6)$ discretization error finite difference approximation to solutions of the Poisson equation in three variables, Dept. of Computer Science Report CSD-TR 230, April 19, 1977, Purdue Univ.
14. R.E. Lynch and J.R. Rice (1978), High accuracy finite difference approximation to solutions of elliptic partial differential equations, Proc. Nat. Acad. of Sci. (to appear).
15. ________________ (1979a), A high order difference method for differential equations, Math. Comp. (to appear); also: Dept. of Computer Science Report CSD-TR 244, September, 1977, Purdue Univ.
16. ________________ (1979b), The Hodie method for second order elliptic problems (to appear).
17. W.E. Milne (1953), Numerical Solution of Differential Equations, John Wiley and Sons.
18. M.R. Osborne (1967), Minimizing truncation error in finite difference approximations to ordinary differential equations, Math. Comp. 21 133-145.
18. J.R. Rice (1977), ELLPACK: A research tool for elliptic partial differential equations software, in Mathematical Software III (J.R. Rice, Edit.) Academic Press, 319-342.
20. J.B. Rosser (1975, Nine point difference solutions for Poisson's equation, Comp. and Maths. with Appls., 1 351-360.

21. D.M. Young and J.H. Dauwalder (1965), Discrete representations of parital differential operators, in Errors in Digital Computation (L. Rall, Edit.), Academic Press, 181-207.

The authors were partially supported by National Science Foundation Grant G 7701408.

Division of Mathematical Sciences
Purdue University
West Lafafayette, Indiana 47907

Solving ODE'S with Discrete Data in SPEAKEASY

L. F. Shampine

1. INTRODUCTION

SPEAKEASY [2] is a very high level language designed to make using computers easy for model building. This paper describes the results of an attempt to provide a capability for solving differential equations in the language. Although SPEAKEASY is a scientific tool seeing significant use, this is not our main reason for the study. The language is representative of situations in which one wants to regard the solution of a differential equation as a high level operation requiring no user intervention at all. This kind of operation is beyond the state of the art because present codes require significant user input, and possibly intervention, and because present codes (with at least the exception [7]) do not even try to estimate error in the sense expected by the unsophisticated user, much less control it. The situation is in part due to considerations of efficiency, it being too expensive to achieve some desirable objectives in general scientific computation. In the context of SPEAKEASY the problems are relatively inexpensive compared to those of general library usage, and the design decision was made to pay any reasonable cost to keep computation as reliable and as simple as possible for the user. Thus the language affords a natural setting in which to find out how far the boundaries of software for ordinary differential equations can be extended.

Without going into the nature of SPEAKEASY the reader

ISBN 0-12-208360-1

will not be able to see the origin of the difficulty, but the fact is that the differential equations to be solved are defined only at certain points. Thus it is necessary to confront the basic question of solving differential equations when it is not possible to evaluate the equation anywhere we like. This is completely different from the task presented library codes, though it is not unusual because of coefficients which are measured data. Investigation of the matter also helps advance the state of the art of solving ordinary differential equations.

The formulas employed in our code were first brought to our attention by J. Barkley Rosser in a talk which subsequently appeared in [4]. We have studied them in the papers [6,9] and in the dissertation of H. A. Watts [8]. Once again we find them useful and it seems most appropriate that we describe their application on this occasion honoring Professor Rosser.

2. THE PROBLEM

SPEAKEASY does not have within it functions in the sense of FORTRAN and other general purpose programming languages. A function like sin(x) can be employed in SPEAKEASY but it is defined as a table of values on a specified set of points and it is held in the computer as an array. For this reason we cannot even talk about the solution of a general problem like $y' = f(x,y)$ because the function f would have to be defined on all the necessary values of y when the problem is posed within the language, and these values are not known. The most general problem that can be readily described in SPEAKEASY is the class of linear problems

$$\underline{y}' = a(x)\underline{y} + \underline{g}(x)$$

where

$$a(x) = \left(a_{ij}(x)\right) \qquad 1 \le i, \quad j \le n$$

$$\underline{g}(x) = \left(g_i(x)\right) \qquad 1 \le i \le n$$

$$y(a) \text{ given, integrate from } x = a \text{ to } x = b,$$

and we shall restrict our attention to them.

When the user defines the problem he provides the matrix a(x) and the vector g(x). To do this he had to specify the argument x. By this action he has implicitly defined where the solution y(x) is to be evaluated. Because the user is unlikely to have good information about where it is appropriate to evaluate y and because equal spacing in x is convenient, we shall suppose this to be the case. Thus we presume that the problem is to be integrated from XBEGIN to XEND using M equally spaced points. This corresponds to a spacing, or step size, H = (XEND-XBEGIN)/(M-1).

Solving the problem is defined in an unusual way. We want solution values on the given mesh as accurate as it is possible to get them. In particular, it is the true, or global, error we wish to control rather than the local error that is the usual object of our attention. It is important that we ask the user to interact with the solution process only if it is unavoidable. It may be presumed that the number of equations and the number of mesh points is not so large as to cause storage difficulties and that any moderately efficient solution procedure will result in acceptable computing times.

3. A SKETCH OF THE ALGORITHM

We shall present an algorithm and a code for realizing it. The code is written in FLECS [1] which is a FORTRAN preprocessor. The output of the preprocessor is a FORTRAN subroutine which can be run in SPEAKEASY. However, the reason for using FLECS is that the control structures available make the workings of the algorithms transparent when coded. This is appropriate for the explanations provided by this paper and is also very convenient for translation to other languages. In addition we used the Basic Linear Algebra Subprograms [3] which are standard FORTRAN modules for performing basic operations such as adding two vectors. These are operations facilitated by SPEAKEASY so that not only do they clarify the coding, they suggest to us that it would be easy to write the subroutine in the SPEAKEASY language itself were this to be deemed desirable.

An important goal of improved programming methodology is

to make a code easy to modify. The work reported here is an enhanced version of that described in the report [5]. A significant algorithmic change was made which had the effect of increasing the length of the code more than 20%. The programming style made it quite clear exactly which parts of the code had to be modified and made it very easy to carry out these changes. The first computer run revealed that an absolute value sign had been neglected but after correcting it, we had a functioning code with which to experiment in the selection of certain parameters. Our experience was that the methodology was extremely helpful in all phases of the project.

FLECS makes describing the code quite easy. Leaving aside the call list, comments, dimension statements, and data statements, the code itself is merely

```
INITIALIZE-COEFFICIENTS-AND-DIVISORS
DO-SWEEPO
IF(MUSED.NE.1)DO-UP-TO-FOUR-MORE-SWEEPS
PREPARE-TO-RETURN
RETURN
```

The statements here involve procedures in FLECS which may be thought of as blocks of code or subroutines. First we initialize the coefficients defining the formulas. They are set via assignment statements because this method makes the indices very clear. Also, the coefficients are all integers or ratios of integers, the largest of which is 65112, so that there should be no trouble getting them into any medium sized computer accurately. Sweep 0 does some initialization and determines how much of the integration is possible with the given step size. MUSED is the number of mesh points actually used in the integration. If there is more than one mesh point, we successively sweep through the mesh generating more and more accurate solution values. When this process yields no more improvement, we gather together information the user needs: We tell him the interval on which the problem was successfully solved and how accurately, and what to do to get a more accurate solution. We also tell him what interval remains to be solved and what action he must take to solve it. Finally we do a FORTRAN RETURN from the subroutine.

In succeeding sections we shall describe the algorithms and their development more fully. The FLECS code is heavily commented and we shall often refer to it to connect up the theory and practice.

4. CHOICE OF A BASIC METHOD

The constraints imposed by the structure of the SPEAKEASY language greatly limit our options for a basic method. The fundamental difficulty that the equation is defined only at discrete points is one of some general importance because it occurs when any function in a differential equation arises from experimental data. There are two distinct approaches possible: One is to fit the data so as to generate functions which we can evaluate anywhere and then solve the approximating problem as accurately as we like using any one of the pieces of mathematical software generally available. The other is to employ a numerical method which uses only the discrete data available. Both approaches require us to address the problem of the limited accuracy inherent in the discretization. It is not clear how to assess this in the first approach. In the second approach, because our data is assumed exact, the discretization error is exactly that with which we are accustomed to dealing. The only difference is that we cannot adjust the mesh to do anything about it. Thus although the first approach is the easier mechanically, it appears to be far more difficult conceptually. Also, because we seek to do a much better job in our special circumstances than a general purpose code would do, the effort of developing a new integration scheme really cannot be avoided.

Although the user could specify in SPEAKEASY a set of points on which he wishes an approximate solution, in general he is unlikely to have available the information and knowledge as to how to do this. Furthermore it is a basic principle that the solution ought to be as simple as possible. We are led to computing a solution at equally spaced points. A natural development here is to advance the solution as far as possible on such a mesh and to tell the user that he must use a finer mesh on the remainder of the interval. This actually accomplishes a crude variation in the mesh if it is absolutely necessary.

The standard initial value techniques are excluded by the circumstances. Runge-Kutta methods require evaluations at off-mesh points so cannot be used. Methods involving memories, such as the Adams methods, have a starting problem. Starting is usually accomplished by either varying the mesh initially or by a Runge-Kutta method, both of which are excluded. Extrapolation methods in their usual form involve a refinement of the mesh, which is excluded, and regarding the given mesh as the finest in a set of refinements means that solution values are not computed at all the mesh points.

In context it is important to get all possible accuracy on the given mesh. We cannot vary the mesh so varying the formula is all that remains in our bag of tools. The only way we really understand for doing this amounts to using procedures of successively higher orders. An appealing procedure is that of difference correction which successively raises the accuracy of solution on a fixed mesh by applying higher order methods. We chose a family of block one-step methods [9] which can be viewed as variants of a number of the procedures mentioned. The solution is advanced by a block of new values at a time, e.g. using the notation $\underline{y}_n \doteq \underline{y}(x_n)$, one might generate $\underline{y}_1, \underline{y}_2$ and then $\underline{y}_3, \underline{y}_4$, etc. The process can be viewed as a special Runge-Kutta procedure which does all its evaluations at mesh points and for which the intermediate computations represent solution approximations of the same order as the final value at the end of the "step." The scheme we employ uses the highest order possible which, as it turns out, can be viewed as arising in the same way as the Adams formulas but differs by simultaneously getting a block of new solution values rather than remembering all but one value from previous computations. Applying these block methods so that the order is increased on successive sweeps of solution is in essence a difference correction procedure. It is our intention to develop the code further to deal with stiff problems, so it is very convenient that these formulas are also suitable for this task. The formulas have the form

$$\underline{y}_{n+i} = \underline{y}_n + h \sum_{j=0}^{b} \alpha_{ij} \left(a(x_{n+j}) \underline{y}_{n+j} + \underline{g}(x_{n+j}) \right) \quad i = 1, \ldots, b \qquad (1)$$

when a block of b new values is generated at a time. The order which can be achieved by the Newton-Cotes family [8, Appendix C] is

block length b	1	2	4	6	8
order of convergence	2	4	6	8	10

and all these formulas are A-stable. It is not likely that very fine meshes will be used in SPEAKEASY because of the storage implications. For this reason very high orders are not likely to be used. Indeed, our experiments showed that the order 10 formula was never used in what seemed to be normal computation. And so we limited our formulas to order eight.

5. NECESSARY CONDITIONS ON THE STEP SIZE

Because the step size is given to the code without knowledge of the behavior of the solution, it is quite possible that it is so large as to cause the computations to be unstable (to "blow up") or to cause our judgements based on asymptotic behavior to fail because they are not applicable. This situation does not arise so seriously in more typical computation because codes which vary their step size do so to maintain stability and to help assure the validity of asymptotic behavior.

Suppose that at x_j there is an error of amount $\underline{\delta}_j$ in the solution. If we were to continue the integration with no further error, we would compute $\underline{u}(x)$ from

$$\underline{u}' = a(x)\underline{u} + \underline{g}(x) \ , \quad \underline{u}(x_j) = \underline{y}(x_j) + \underline{\delta}_j \ .$$

The difference between $\underline{u}$ and the true solution, $\underline{e}(x) = \underline{u}(x) - \underline{y}(x)$, satisfies

$$\underline{e}' = a(x)\underline{e} \ , \quad \underline{e}(x_j) = \underline{\delta}_j \ .$$

Thus, expanding $\underline{e}(x_j+h)$ in terms of the step size h, we see that

$$\begin{aligned} \underline{e}(x_j+h) &= \underline{e}(x_j) + h\underline{e}'(x_j) + 0(h^2) \\ &= \underline{\delta}_j + ha(x_j)\underline{\delta}_j + 0(h^2) \ . \end{aligned}$$

Clearly the propagation of the error is unstable unless $ha(x_j)\underline{\delta}_j$ is small in some sense. A minimal requirement would appear to be something like

$$\|ha(x_j)\| < 0.1 \tag{2}$$

for some matrix norm. It can happen that the matrix a implies that the error $\underline{\delta}_j$ will be strongly damped but that $\|a\|$ is very large. For such problems, termed stiff problems, the requirement (2) can be unrealistically severe. The code described herein is aimed at non-stiff problems for which $\|a\|$ is of moderate size and we shall consider stiff problems at another time.

We have already stated in (1) the form of the formulas which are to be the basis of our computation. It is our intention to solve for the $\underline{y}_{n+i}$ by simple iteration, i.e. given $\underline{y}_{n+i,0}$ we iterate by

$$\underline{y}_{n+i,m+1} = \underline{y}_n + h \sum_{j=0}^{b} \alpha_{ij}\left(a(x_{n+j})\underline{y}_{n+j,m} + \underline{g}(x_{n+j})\right) \quad i = 1,\ldots,b .$$

If we let the iteration error be $\underline{e}_{i,m} = \underline{y}_{n+i,m} - \underline{y}_{n+i}$ and define

$$|e_m| = \max_{1 \le i \le b} \|e_{i,m}\| , \quad |a| = \max_{x_i} \|a(x_i)\| ,$$

$$\alpha = \max_{1 \le i \le b} \sum_{j=1}^{n} |\alpha_{ij}|$$

for suitable vector and matrix norms, then it is easy to see that

$$|e_{m+1}| \le |e_m| \; |h| \; \alpha \; |a| .$$

The error of iteration will tend to zero and the method will be well defined if $|h| \, \alpha \, |a| < 1$.

For the methods considered $\alpha < 5.72$ so that if we require $12|h||a| < 1$, we are assured that the numerical methods are well defined, the error will not propagate too fast, and the iteration error is reduced by at least half each iteration. This test will be applied to see how far we might reasonably _try_ to integrate and to deduce how many equally spaced mesh points appear necessary for the integration of the remainder of the interval. A matrix norm which is cheap to compute is

$$\|a(x)\| = \max_{j=1,\ldots,n} \sum_{i=1}^{n} |a_{ij}(x)|$$

and we use it in the test.

The procedure DO-SWEEPO does some initialization first -- it computes the step size and defines the initial solution to be a constant over the interval. Then it runs through the interval testing if $12|h||a| < 1$. The integration will proceed from XBEGIN as far towards XEND as this condition is satisfied, namely XQUIT. The formulas require at least 7 points so if the condition is not satisfied for this many points we cannot integrate at all and must set XQUIT=XBEGIN. MUSED is the number of points actually used, that is, the number of mesh points in the interval from XBEGIN to XQUIT. If XQUIT≠XEND, we must run through the remainder of the interval computing $12|a|$ so that we can tell the user how fine a mesh he must use if he is to integrate from XQUIT to XEND. The actual computation of the necessary number of mesh points is done in the procedure PREPARE-TO-RETURN.

6. VARYING THE ORDER

The basic idea for computing solutions as accurately as possible on the given mesh and for estimating the global error is to compute independently solutions of increasingly high order and compare them. We start off with a solution of order 0 (a constant solution) from the initialization in sweep 0. Then in sweep L we carry out the integration using a formula of order H^{2L} based on a block involving max(2,2L-1) points. We compare the solution at each mesh point to that of the previous sweep, which is of order $H^{2(L-1)}$. Our hypothesis is that the higher order solution is more accurate until we find evidence that this is not true. Certainly for "small" step sizes the higher order result is much more accurate and provides an excellent estimate of the global error E_{L-1} of the result of sweep(L-1). We chose to increase the order of the formulas by two at each sweep to improve the quality of the estimate as well as to get to high order results quickly.

In sweep L we form E_{L-1}. If L = 1, we automatically raise the order and go to L = 2. If L > 1, we compare E_{L-1} to E_{L-2}. If $E_{L-1} \leq E_{L-2}$, we conclude that raising the order is improving the solution and we raise the order unless we have already reached the maximum of L = 4. If $E_{L-1} > E_{L-2}$ or if L = 4, we stop raising the order and accept the result of order $H^{2(L-2)}$ or H^6 as our answer, respectively. As we have previously noted, for practical meshes there seemed to be no point in doing more than four sweeps.

The procedure DO-UP-TO-FOUR-MORE-SWEEPS carries out the algorithm just outlined. It calls upon a procedure to generate a new solution of order H^{2L}. Then it calls upon a procedure to estimate the global error of the result of the preceding sweep. This is done by comparison as we have said but the details will be discussed in the next section. Finally it is decided whether or not to raise the order by doing another sweep. The basic linear algebra subprogram SCOPY is used to swap solution arrays so as to retain only those needed for the tests in order to hold down storage used.

Generating a solution is done in the procedure GENERATE-A-SOLUTION and is straightforward with one exception. We are using block methods so that there is a problem at the end of the integration. What we do is to discard some solution values from the next to last block so that we can end up at XQUIT. We might remark that this is the analog of the starting problem for methods involving memory but we have a far simpler solution for this special class of formulas.

The computation of a block of new solution values is done in the procedure COMPUTE-A-BLOCK. The initial guesses are taken from the previous sweep so have a global accuracy of order $H^{2(L-1)}$. The formulas being used in sweep L have a local error of order H^{2L+1}. It is easy to see that each iteration increases the local accuracy by one order so that we insist on at least three iterations to get the correct order. Iterations are relatively cheap in this context. Furthermore, at low orders when the guessed values are poor, α is small and the rate of convergence is high. At high orders the guessed values are likely to be quite good and the rate of convergence is never worse than 0.5. We have not encountered a problem in our limited testing which needed more than an average of 4 iterations to achieve convergence so we limited the process to 20 iterations. Because we are guaranteed a rate of convergence of at least 0.5, a standard result about contraction mappings says that the norm of the difference of successive iterates bounds the error in the last iterate. To preserve significance the iteration is framed in terms of computing the difference between iterates. In addition, if the norm of the difference does not decrease, we know that we have reduced it as much as possible in the precision being used and we must terminate the iteration. In general we terminate when the last iterate in sweep L has no more error than 10^{-3L}. This choice is heuristic with the goal of not wasting time computing very accurately a crude approximation yet computing it accurately enough that raising the order improves the overall accuracy if asymptotic approximations are valid.

By iterating until convergence is achieved we preserve the stability properties of the formulas employed. However the <u>way</u> we iterate imposes a step size restriction entirely analogous to that arising from a formula with a finite stability region. If we were to alter the iteration so as to use some variant of Newton's method, we could take advantage of the A-stability of the formulas and so solve stiff problems. There are a number of difficulties in doing this so we have deferred the solution of stiff problems to another time.

7. ERROR MEASUREMENT

To decide whether to raise the order and, at the end of the computation, to assess the merit of the solution, we must somehow measure the error. The measurement must be simple enough that we can convey easily to the user the quality of the solution and it must be realistic and safe. A pure absolute error measure is not realistic because it does not reflect the scale of the solution. A pure relative error takes scale into account but is unsafe at a zero of the solution. We have chosen to measure the error in an absolute sense if the solution is less than one in magnitude and in a relative sense if the solution is greater than one in magnitude. We are concerned about the error throughout the range of integration so it is the worst error that we estimate. To keep matters simple we also consider only the worst error in any equation. The error in the $O(H^{2(L-1)})$ solution of equation I is measured by comparing the solution to that of the $O(H^{2L})$ solution. Thus

$$E_{L-1} = \max_{\text{K in interval}} \ \max_{\text{eqn I}} \frac{\left|\left(O(H^{2L})\text{sol'n} - O\left(H^{2(L-1)}\right)\text{sol'n}\right)_{\text{eqn I}}\right|}{\max\{1, |O(H^{2L})\text{sol'n eqn I}|\}}$$

which is computed in the procedure COMPUTE-GLOBAL-ERROR.

If the user of the code is dissatisfied with the estimated maximum global error, he will have to solve the problem again with a smaller step size. But how much smaller? We chose to return to him a factor R such that if he uses R times as many mesh points in a new computation, he should get about one order of magnitude more accuracy.

Suppose that the accepted solution is of order 2L so that the error is (asymptotically) proportional to H^{2L}. We wish to increase the number of mesh points by a factor R, which is equivalent to reducing H by a factor of 1/R, so that the error is reduced by a factor of 0.1. Thus

$$R = \frac{1}{(0.1)^{\frac{1}{2L}}} = 10^{\frac{1}{2L}} .$$

Unfortunately it is possible that the solution accepted be of order 0, meaning that the step size is not small enough for asymptotic results to be relevant. In such a case we have no guide for choosing a factor R. We arbitrarily take R = 10 in such a case. Notice that if the order were two, the factor would be about 2 so this is not way out of line if we should go to order two in the next computation by virtue of having attained a small enough step size that asymptotic results are meaningful, yet it is large enough that we can rapidly get into this range. The computation of R is done in the procedure PREPARE-TO-RETURN.

The R selected in this way may be quite conservative

because it is based on the order used in the given computation. As a general rule, if the step size is small enough, a high order method will be more accurate than a lower order one. Thus as the mesh is refined the code will tend to go to higher orders. If an R is selected at one order and in a subsequent computation the code goes to a higher order, we expect that the error will be reduced by more than a factor of 10.

8. SOME EXAMPLES

Our first example will clarify the role of the preliminary tests on the mesh made in sweep 0 of the code. Consider the problem

$$\underline{y}' = \begin{pmatrix} -10 & 0 \\ 0 & -20 \end{pmatrix} \underline{y} + \begin{pmatrix} e^{-10x} \\ xe^{-20x} \end{pmatrix} ,$$

$$\underline{y}(0) = \begin{pmatrix} 1 \\ 1 \end{pmatrix} , \quad \text{integrate from 0 to 10} .$$

This is a simple example of a stiff problem, meaning here that an unrealistically small step size would be necessary to solve the problem with a code like ours -- although it will do so if you are willing to use this step size. When the code was called with eleven mesh points, it returned with the information that XQUIT=XBEGIN=0 and MUSED=1, meaning that it did not advance the solution at all, and with MMORE=2401, meaning that at least 2401 mesh points would be necessary to integrate this problem. A "large" value of MMORE is the way that stiffness will be reported. However, there are other conditions which could result in large values returned in this parameter. It is presumed that the solution components are smooth. It is not possible for the code to be presented a problem in SPEAKEASY which exhibits a discontinuity or singularity. This is because the code has only the finite values given at mesh points to work with. It cannot distinguish between problems which are extremely smooth between mesh points and those becoming infinite there; it presumes they are smooth. It may be that the problem changes abruptly enough that this is reflected in the discrete data and causes the code to return a large value of MMORE. A much sharper indication of a lack of smoothness is the parameter R and we shall discuss it in a moment.

The next example might be termed a normal one. We have

$$\underline{y}' = \begin{pmatrix} 0 & -2x \\ 2x & 0 \end{pmatrix} \underline{y} + \begin{pmatrix} 2x \\ 0 \end{pmatrix},$$

$$\underline{y}(0) = \begin{pmatrix} 1 \\ 1 \end{pmatrix}, \quad \text{integrate from 0 to 2}$$

for which the solution is

$$\underline{y}(x) = \begin{pmatrix} \cos(x^2) \\ 1 + \sin(x^2) \end{pmatrix} .$$

As x increases the solution components oscillate increasingly rapidly so that with a constant mesh spacing we should expect the solution to become increasingly difficult. When the code was called with 41 mesh points it returned with the result that it had successfully integrated from 0 to XQUIT=0.8 using 17 points of the original mesh. The worst global error was estimated to be 4.2×10^{-6} in the measure discussed in the preceding section. When we computed the true global error by comparison with the known solution we found that the worst error was 4.1×10^{-6}. To reduce the error by a factor of 10 the code estimated that the number of mesh points would need to be increased by a factor R = 1.78. The code also returns information about the solution of the remainder of the interval. It said that at least 58 points ought to be used. When we solved the remaining interval of 0.8 to 2 with 59 points, the computation succeeded. The estimated global error was 2.7×10^{-6} and the true global error was 3.6×10^{-6}.

The third problem involves only one equation and integrates in the direction of decreasing x.

$$y' = \frac{-8(x-1)^7}{1+(x-1)^8}\, y \quad ,$$

$$y(2) = 5, \text{ integrate from 2 to 0 .}$$

The solution

$$y(x) = \frac{10}{1+(x-1)^8}$$

has a peak at x = 1. We first called the code with too few mesh points and it promptly stated that at least 97 mesh points were needed. We tried 101, which was successful. The solution was computed over the entire interval with an estimated error of 4.5×10^{-6}. The true error was 4.4×10^{-6}. The code said that if the number of mesh points were increased by a factor of 1.78, we should reduce the error by a factor of about 10. So, we solved the problem again with 180 mesh points. This resulted in a solution with an estimated error of 6.5×10^{-7} and a true error of 6.5×10^{-7}. In this case the prediction about the error reduction worked rather well. It should if the asymptotic results are valid. If the estimated error were not improved by anything like a factor of 10, this would be good evidence that the asymptotic approximations are invalid because the problem is not smooth, e.g. a discontinuity or singularity is present, or that we have reached the limits imposed by the number of digits available with our computer.

REFERENCES

1. T. Beyer (1974), FLECS -- Fortran language with extended control constructs, User's manual, University of Oregon Computing Center, Eugene, Oregon.

2. S. Cohen and S. C. Pieper (1976), The SPEAKEASY-3 reference manual, Argonne National Laboratory Rept. ANL-8000 Rev. 1, Argonne, Illinois.

3. C. L. Lawson, R. J. Hanson, D. R. Kincaid, and F. T. Krogh (1978), Basic linear algebra subprograms for FORTRAN usage, ACM Trans. Math. Software, (to appear).

4. J. B. Rosser (1967), a Runge-Kutta for all seasons, SIAM Rev., 9, pp. 417-452.

5. L. F. Shampine (1977), Solving ordinary differential equations in SPEAKEASY, Sandia Laboratories Rept. SAND77-1129, Albuquerque, New Mexico.

6. ________________ and H. A. Watts (1969), Block implicit one-step methods, Math. Comp., 23, pp. 730-740.

7. ________________ (1976), Global error estimation for ordinary differential equations, ACM Trans. Math. Software, 2, pp. 172-186.

8. H. A. Watts (1971), A-stable block implicit one-step methods, Sandia Laboratories Rept. SC-RR-71 0296, Albuquerque, New Mexico.

9. ________________ and L. F. Shampine (1972), A-stable block implicit one-step methods, BIT, 12, pp. 252-266.

This work was supported by the U.S. Department of Energy (DOE) under contract no. AT(29-1)-789 and has been authored by a contractor of the United States Government under contract. Accordingly the United States Government retains a non-exclusive, royalty-free license to publish or reproduce the published form of this contribution, or allow others to do so, for United States Government purposes.

Applied Mathematics Research Department
Sandia Laboratories
Albuquerque, New Mexico 87185

Perturbation Theory for the Generalized Eigenvalue Problem

G. W. Stewart

1. Introduction.

In this paper we shall be concerned with the generalized eigenvalue problem

$$Ax = \lambda Bx \tag{1.1}$$

where A and B are complex matrices of order n. The nontrivial solutions x of (1.1) are called eigenvectors and the corresponding values of λ are called eigenvalues. We shall be interested in what happens to the eigenvalues of (1.1) when A and B are perturbed. Specifically let

$$\tilde{A} = A + E \ , \quad \tilde{B} = B + F$$

and consider the generalized eigenvalue problem

$$\tilde{A}\tilde{x} = \tilde{\lambda}\tilde{B}\tilde{x} \ . \tag{1.2}$$

If we can associate with an eigenvalue λ of (1.1) an eigenvalue $\tilde{\lambda}$ of (1.2) that approaches λ as E and F approach zero, then we may ask for bounds on some measure of the distance between λ and $\tilde{\lambda}$ in terms of the size of E and F. We may ask for similar bounds for the eigenvectors x and $\tilde{x}$; however, we shall not consider this more difficult problem in this paper.

Throughout the paper we shall assume that the reader is familiar with the algebraic theory of the generalized eigenvalue problem. The symbol $\|\cdot\|$ will denote the usual Euclidean vector norm or the spectral matrix norm defined by

$$\|A\| = \sup_{\|x\|=1} \|Ax\| \ .$$

ISBN 0-12-208360-1

In the next section we shall consider the drawbacks of an approach to the problem that can be applied whenever A or B is nonsingular: namely, to reduce the generalized eigenvalue problem to an ordinary eigenvalue problem. In §3 we shall give two examples that illustrate the difficulties we may expect to encounter when A and B are singular or nearly so. In §4 we describe the first order perturbation theory for a simple eigenvalue and introduce the chordal metric as a measure of the distance between λ and $\tilde{\lambda}$. This leads directly to a condition number for simple eigenvalues. In §5 we place the first order theory on a rigorous basis by means of a generalized Gerschgorin theory. In §6 we consider a class of Hermitian problems for which the classical min-max theory for the Hermitian eigenvalue problem can be generalized.

Because this is an expository paper, most of the results will be stated without proof. However, §7 will be devoted to a brief bibliographical survey so that the interested reader may pursue the subject further.

2. Reduction to an ordinary eigenvalue problem.

When B is nonsingular we can write (1.1) in the form

$$B^{-1}Ax = \lambda x ,$$

which exhibits λ and x as an eigenvalue and eigenvector of an ordinary problem. Consequently we can apply the highly developed perturbation theory for the ordinary eigenvalue problem. However, this approach has some limitations.

The first limitation is that we must restrict the perturbation F of B so that $\tilde{B}$ is nonsingular. What this means can best be expressed in terms of the condition number

$$\kappa(B) = \|B\| \, \|B^{-1}\|$$

of B with respect to inversion. Unless F satisfies

$$\kappa(B)\frac{\|F\|}{\|B\|} < 1 ,$$

it cannot be guaranteed that B + F is nonsingular. Consequently, if B is "ill-conditioned" in the sense that $\kappa(B)$ is large, we must restrict ourselves to small perturbations F .

The second limitation is that the perturbation in $B^{-1}A$ may be large, even though the perturbations in A and B are small. Let

$$\tilde{B}^{-1}\tilde{A} = B^{-1}A + H .$$

Then it can be shown that, up to second order terms in E and F ,

$$\frac{\|H\|}{\|B^{-1}A\|} \leq \kappa(B)\frac{\|F\|}{\|B\|} + \frac{\|B^{-1}\| \, \|A\|}{\|B^{-1}A\|}\frac{\|E\|}{\|A\|} .$$

Again if $\kappa(B)$ is large, $\|H\|$ can be large relative to $\|B^{-1}A\|$, and the perturbation theory will predict large perturbations for the eigenvalues, even the ones that are insensitive to small perturbations in A and B.

Thus when B is ill conditioned, working with $B^{-1}A$ is unlikely to yield very satisfactory results. However, if B is well conditioned, one will get fairly sharp bounds from the approach. Where to draw the dividing line must be decided individually for each problem.

In the important case where A is Hermitian and B is positive definite, it is customary to reduce (1.1) to a Hermitian eigenvalue problem as follows. The matrix B has a Cholesky factorization of the form

$$B = R^H R ,$$

where R is upper triangular. If we set $y = R^{-1}x$, then (1.1) can be written

$$(R^{-H}AR^{-1})y = \lambda y , \tag{2.1}$$

and one can apply perturbation theory for the Hermitian eigenvalue problem to (2.1). However, the approach has the same drawbacks sketched above.

3. Two examples.

In this section we shall give two examples. The first illustrates the fact that the usual notion of an ill-conditioned eigenvalue cannot be transfered directly to the generalized eigenvalue problem. The second illustrates the pathology of a truly ill-conditioned eigenvalue.

For the first example consider the matrices

$$A = \begin{pmatrix} 0 & 0 \\ 0 & 1 \end{pmatrix} \qquad B = \begin{pmatrix} 1 & 0 \\ 0 & 0 \end{pmatrix} .$$

The associated generalized eigenvalue problem has an eigenvalue $\lambda_1 = 0$ corresponding to the vector $(1,0)^T$. It may also be said to have an infinite eigenvalue λ_2 corresponding to the vector $(0,1)^T$, since

$$\frac{1}{\infty}A\begin{pmatrix} 0 \\ 1 \end{pmatrix} = B\begin{pmatrix} 0 \\ 1 \end{pmatrix} .$$

Now if

$$\varepsilon = (\|E^2\| + \|F\|^2)^{\frac{1}{2}} , \tag{3.1}$$

Then it follows from results to be presented later that

$$|\lambda_1 - \tilde{\lambda}_1| = O(\varepsilon) .$$

On the other hand, since λ_2 will in general be finite, the best statement we can make about the difference between λ_2 and $\tilde{\lambda}_2$ is that

$$|\lambda_2 - \tilde{\lambda}_2| \leq \infty .$$

Since small changes in A and B make large ones in λ_2, one is tempted to conclude that λ_2 is ill conditioned, in the sense that the term is used by numerical analysts. However, observe that λ_2 corresponds to an eigenvalue $\mu_2 = 0$ of the reciprocal problem

$$Bx = \mu Ax . \tag{3.2}$$

With the roles of A and B thus reversed, we can say that

$$|\lambda_2^{-1} - \tilde{\lambda}_2^{-1}| \equiv |\mu_2 - \tilde{\mu}_2| = O(\varepsilon) .$$

Hence, although λ_2 varies violently with small perturbations in A and B, its reciprocal is well behaved. The apparent ill-conditioning of λ_2 is in fact a consequence of the way we have chosen to measure distance around the point at infinity. In the next section we shall show how the use of the chordal metric circumvents this difficulty.

For the second example let

$$A = \begin{pmatrix} 1 & 0 \\ 0 & 1/2\times 10^{-8} \end{pmatrix}, \quad B = \begin{pmatrix} 1 & 0 \\ 0 & 10^{-8} \end{pmatrix} .$$

The two eigenvalues of this problem are $\lambda_1 = 1$ and $\lambda_2 = 1/2$, again corresponding to the vectors $(1,0)^T$ and $(0,1)^T$. Although these appear to be reasonable numbers, note that even if we restrict ε to be less than $2\cdot 10^{-8}$, $\tilde{\lambda}_2$ can be made to assume any value whatsoever. Thus λ_2 must be considered to be truly ill-conditioned.

Unfortunately, λ_1 can inherit some of the ill-conditioning of λ_2. Specifically, let

$$\tilde{A} = \begin{pmatrix} 1 & 10^{-8} \\ 10^{-8} & 10^{-8} \end{pmatrix}, \quad \tilde{B} = B \tag{3.3}$$

so that $\varepsilon < 2\cdot 10^{-8}$. Then it is easily verified that the eigenvalues of the perturbed problem are 1 ± 10^{-4}. Thus a perturbation of order 10^{-8} in A has induced a perturbation of order 10^{-4} in λ_1.

The etiology of this behavior may be described informally as follows. The matrices A and B are very near the matrices $\hat{A} = \hat{B} = \text{diag}(1,0)$, which have the common null vector $\hat{x}_2 = (0,1)^T$. Any number whatsoever is an eigenvalue corresponding to $\hat{x}_2$, since the relation (1.1) is satisfied for any value of λ. On the other hand, there is a unique eigenvalue

$\hat{\lambda}_1 = 1$, corresponding to $\hat{x}_1 = (1,0)^T$. Now perturbations in $\hat{A}$ and B will in general cause an eigenvalue $\tilde{\lambda}_2$ to coalesce. If $\tilde{\lambda}_2$ emerges near $\hat{\lambda}_1$, then the latter can be greatly perturbed.

Thus we see that problems in which A and B have nearly common null spaces give rise to unstable eigenvalues that can affect otherwise stable eigenvalues, although in fact it is unlikely that they will. This places an unfortunate limitation on what we can say a priori about the generalized eigenvalue problem; in general we shall have to restrict the size of our perturbations so that none of the eigenvalues can move too much. However, it should be possible to prove strong conditional theorems about individual eigenvalues under the assumption that the perturbations leave them well separated from their neighbors.

Incidentally, ill-conditioned problems of this kind can easily be generated in practice. For example, if A and B are Rayleigh-Ritz approximations to two operators that have been obtained by using a nearly degenerate basis, then A and B will have an approximate null vector whose components are the coefficients of a linear combination of the basis that approximates zero.

4. <u>First order theory</u>.

If λ is a simple eigenvalue of (1.1), then for ε small enough $\tilde{\lambda}$ is a differentiable function of the elements of E and F , and consequently we can obtain a simple expression for $\tilde{\lambda}$ that is accurate to terms of order ε^2 . In order to do this we now assume that the eigenvector x has been normalized so that

$$\|x\| = 1 \ ,$$

and we introduce the left eigenvector y that satisfies

$$y^H A = \lambda y^H B \ , \quad \|y\| = 1 \ .$$

Then it is easily seen that if we set

$$\alpha = y^H Ax \ , \qquad \beta = y^H Bx \ ,$$

we have

$$\lambda = \alpha/\beta \ . \tag{4.1}$$

We shall call the numbers α and β the <u>Rayleigh components</u> of λ and, as is customary, call the expression (4.1) a Rayleigh quotient.

The results of the first order perturbation theory may be summarized in the equation

$$\tilde{\lambda} = \frac{\alpha + y^H Ex + O(\varepsilon^2)}{\beta + y^H Fx + O(\varepsilon^2)} = \frac{\alpha' + O(\varepsilon^2)}{\beta' + O(\varepsilon^2)} \ . \tag{4.2}$$

In this expression we may already discern what makes an eigenvalue ill-conditioned. Namely if α and β are both small, the values of $y^H Ex$ and $y^H Fx$ may overwhelm them. In this case, by suitably choosing E and F we may cause the ratio α'/β' to assume any value at all.

This observation may be made precise by casting the relation between λ and $\tilde{\lambda}$ in terms of the chodal metric defined by

$$\chi(\lambda,\tilde{\lambda}) = \frac{|\lambda-\tilde{\lambda}|}{\sqrt{1+\lambda^2}\,\sqrt{1+\tilde{\lambda}^2}} .$$

Geometrically, $\chi(\lambda,\tilde{\lambda})$ is half the length of the chord connecting λ and $\tilde{\lambda}$ when they have been projected in the usual way onto the Riemann sphere. It is bounded by one and is well defined at infinity.

A short computation shows that (4.2) implies that

$$\chi(\lambda,\tilde{\lambda}) \leq \frac{\varepsilon}{\gamma} + O(\varepsilon^2) ,$$

where ε is defined as usual by (3.1) and

$$\gamma = \sqrt{\alpha^2+\beta^2} .$$

The number γ^{-1} serves as a condition number for λ in the sense that it measures how a perturbation of size ε in A and B affects λ. It is large precisely when the Rayleigh components are small.

It is instructive to compute values of γ for the two examples of the last section. In the first example $\gamma=1$ for both eigenvalues, which supports our assertion that both should be considered well conditioned. In the second example the first eigenvalue has $\gamma=1$ while the second has

$$\gamma = \frac{\sqrt{5}}{2} \cdot 10^{-8} .$$

Thus the second eigenvalue is violently ill-conditioned.

There remains the anomaly that the supposedly well-conditioned eigenvalue of one in the second example is badly perturbed by the perturbation (3.3) in A. The anomaly may be resolved by observing that, owing to the extreme ill-conditioning λ_2, the perturbation in A is of a size for which the first order perturbation theory breaks down. This shows the necessity of following up a first order analysis with rigorous bounds and domains of applicability. For the ordinary eigenvalue problem this is done in two ways, depending on whether the matrix is Hermitian or not. For non-Hermitian matrices one applies elaborations of the Gerschgorin theorem. For Hermitian matrices one applies the classical min-max theory to obtain multiplicity independent bounds. These approaches can be extended to the generalized eigenvalue problem, and in the next two sections we describe the resulting perturbation theory.

5. Gerschgorin theory.

The basis for the theory of this section is the following generalization of the well-known Gerschgorin theorem. To state it, we introduce the vectors

$$a_i^H = (a_{i1}, \ldots, a_{i,i-1}, a_{i,i+1}, \ldots, a_{in})$$

and

$$b_i^H = (b_{i1}, \ldots, b_{i,i-1}, b_{i,i+1}, \ldots, b_{in}) ,$$

which are simply the rows of A and B with their diagonal elements removed.

Theorem 5.1. Let

$$G_i = \left\{ \frac{a_{ii} + a_i^H w}{b_{ii} + b_i^H w} : \|w\|_\infty \leq 1 \right\} \tag{5.1}$$

(here $\|w\|_\infty = \max\{|w_i|\}$). If λ is an eigenvalue of (1.1), then

$$\lambda \in \bigcup_{i=1}^{n} G_i .$$

Thus the theorem states that every eigenvalue of (1.1) belongs to at least one of the regions defined by (5.1). When $B = I$ the regions G_i reduce to the usual Gerschgorin disks

$$\{\lambda : |\lambda - a_{ii}| \leq \sum_{j \neq i} |a_{ij}|\} .$$

The ability of the Gerschgorin theorem to provide precise information about the location of eigenvalues is due to the fact that if a set of k disks are isolated from the others, then that set contains exactly k eigenvalues. In particular, if a single disk is isolated from the others, it contains a single eigenvalue, and if the radius of the disk is small, it gives a tight estimate of the location of the eigenvalue. These properties are shared by the Gerschgorin regions.

Theorem 5.2. If the union K of k of the Gerschgorin regions G_i is disjoint from the remaining regions, then K contains exactly k eigenvalues.

The regions G_i are rather complicated to work with. However, by enlarging them we can replace them with disks on the Riemann sphere. Specifically, let

$$\hat{a}_{ii} = \max \{0, |a_{ii}| - \|a_i\|_1\}$$

and

$$\hat{b}_{ii} = \max\{0, |b_{ii}| - \|b_i\|_1\},$$

where $\|x\|_1 = \Sigma|x_i|$. Set

$$\rho_i = \frac{\|a_{ii}b_i^H - b_{ii}a_i^H\|_1}{\sqrt{|a_{ii}^2| + |b_{ii}^2|}\sqrt{\hat{a}_{ii}^2 + \hat{b}_{ii}^2}}.$$

Then an easy calculation shows that

$$G_i \subset \hat{G}_i = \{\lambda : \chi(a_{ii}/b_{ii}, \lambda) \le \rho_i\}.$$

We shall now show how the Gerschgorin theorem can be used to derive rigorous perturbation bounds. To simplify the exposition we illustrate the technique for a simple eigenvalue of a 2×2 problem. Let $X = (x_1, x_2)$ and $Y = (y_1, y_2)$ be the matrices of right and left eigenvectors of the problem. Then

$$Y^H\tilde{A}X = \begin{pmatrix} y_1^HAx_1 + y^HEx_1 & y_1^HEx_2 \\ y_2^HEx_1 & y_2^HAx_2 + y_2^HEx_2 \end{pmatrix}$$

$$= \begin{pmatrix} \alpha_1' & \varepsilon_{12} \\ \varepsilon_{21} & \alpha_2' \end{pmatrix},$$

and likewise

$$Y^H\tilde{B}X = \begin{pmatrix} \beta_1' & \phi_{12} \\ \phi_{21} & \beta_2' \end{pmatrix}.$$

Since the columns of X and Y have norm unity

$$|\varepsilon_{ij}|, |\phi_{ij}| \le \varepsilon, \tag{5.2}$$

where ε is defined by (3.1).

Now the α_i' and β_i' are the same as in equation (4.2). Hence the center of the Gerschgorin disks for $Y^H\tilde{A}X$ and $Y^H\tilde{B}X$ are the first order approximations to the perturbed eigenvalues. Unfortunately it follows from (5.2) that the radii of the disks are of order ε instead of ε^2, which is what we require if we are to make (4.2) rigorous. However, we can reduce the radii of the disk surrounding λ_1 by a scaling strategy analogous to the method of diagonal similarities that is used with the ordinary eigenvalue problem.

Consider the matrices

$$\begin{pmatrix} \alpha_1' & \tau\varepsilon_{12} \\ \varepsilon_{21} & \tau\alpha_2' \end{pmatrix} , \begin{pmatrix} \beta_1' & \tau\phi_{12} \\ \phi_{21} & \tau\beta_2' \end{pmatrix} .$$

The centers of the Gerschgorin disks for these matrices do not change with τ ; however, their radii do, and an elementary calculation gives the following bounds:

$$\rho_1(\tau) \leq \frac{\sqrt{2}\,\varepsilon\tau}{\gamma_1' - \sqrt{2}\,\varepsilon\tau} ,$$

$$\rho_2(\tau) \leq \frac{\sqrt{2}\,\varepsilon}{\tau\gamma_2' - \sqrt{2}\,\varepsilon} ,$$

where

$$\gamma_i' = \sqrt{\alpha_i'^2 + \beta_i'^2} .$$

Now as τ is decreased, $\rho_1(\tau)$ decreases and $\rho_2(\tau)$ increases. We shall try to reduce ρ_1 as much as possible while keeping the two disks separated.

Let

$$\delta = \chi(\alpha_1'/\beta_1', \alpha_2'/\beta_2') \leq 1 .$$

We first wish to restrict $\rho_2(\tau)$ so that $\rho_2(\tau) \leq \delta/2$. This will certainly be true if we take

$$\tau = \frac{(\sqrt{2} + 1/\sqrt{2})\varepsilon}{\gamma_2'\delta} . \tag{5.3}$$

We now assume that ε is small enough so that $\tau \leq 1$. Then

$$\rho_1(\tau) \leq \frac{3\varepsilon^2}{\gamma_2'\delta(\gamma_1 - \sqrt{2}\varepsilon)} . \tag{5.4}$$

As ε approaches zero we must ultimately have $\rho_1(\tau) < \delta/2$, and $\tilde{\lambda}_1$ will be isolated in G_1 . Moreover the radius of G_1 approaches zero as ε^2 , which justifies (4.2).

This process is quite general, and may be used to obtain bounds for any simple eigenvalue. It follows from (5.4) and the requirement that $\rho_1(\tau) \leq \delta/2$, that the domain of applicability of the theory depends on γ_2' , which for small ε is essentially the reciprocal of the condition number of λ_2 . This quantifies our early assertion that ill conditioning in one eigenvalue can have a deleterious effect on the bounds for the others.

6. Min-max theory.

In Section 2 we saw that when A is Hermitian and B is positive definite, the generalized eigenvalue problem can be reduced to a symmetric eigenvalue problem. In this case the eigenvalues of the problem are real, and if X is the matrix of eigenvectors of the problem, then X^HAX and X^HBX are diagonal. However, the requirement that B be positive definite is too restrictive. For example, if there are constants c and s such that

$$\hat{B} = cB - sA \tag{6.1}$$

is positive definite, then the problem (1.1) can be reduced to the equivalent problem

$$\hat{A}x = \hat{\lambda}\hat{B}x\ , \tag{6.2}$$

where

$$\hat{A} = cA + sB \tag{6.3}$$

and

$$\hat{\lambda} = \frac{s + c\lambda}{c - s\lambda}\ .$$

There is a class of problems for which the transformation (6.1) exists. Let

$$\gamma(A,B) = \inf_{\|x\|=1} \{(x^HAx)^2 + (x^HBx)^2\}^{\frac{1}{2}}\ .$$

A Hermitian generalized eigenvalue problem will be called definite if $\gamma(A,B) > 0$. The basic result about definite problems is contained in the following theorem.

Theorem 6.1. If $\gamma(A,B) > 0$, then there are numbers c and s such that $\hat{B}$ as defined by (6.1) is positive definite.

In outline the proof goes as follows.

It can be shown that the generalized field of values

$$V(A,B) = \{(x^HAx,\ x^HBx) : \|x\| = 1\}$$

is a compact, convex set. The condition $\gamma(A,B) > 0$ says that V cannot contain the origin. Hence it can be rotated counter clockwise through an angle ϕ so that it lies in the upper half plane. The resulting set is $V(\hat{A},\hat{B})$ where $c = \cos\phi$ and $s = \sin\phi$ in (6.1) and (6.3) . But if $V(\hat{A},\hat{B})$ lies in the upper half plane, then $x^H\hat{B}x > 0$ for all $x \neq 0$, and $\hat{B}$ is positive definite.

The matrix of eigenvectors of the problem (6.2) diagonalizes $\hat{A}$ and $\hat{B}$. Thus a definite generalized eigenvalue problem has real eigenvalues and a set of linearly independent eigenvectors that diagonalize A and B .

In the perturbation theorem for the definite generalized eigenvalue problem it will be necessary to restrict E and F so that the perturbed problem is also definite. Fortunately $\gamma(A,B)$ satisfies

$$\gamma(A + E,\ B + F) \geq (A,B) - \varepsilon\ ,$$

where ε is defined as usual by (3.1). Hence if

$$\varepsilon < \gamma(A,B)\ ,$$

then $\gamma(\tilde{A},\tilde{B}) > 0$.

The perturbation bounds are most conveniently phrased in terms of angles θ associated with the eigenvalues. Essentially, these are related to the eigenvalues by $\lambda = \tan(\theta - \text{const.})$; however, some care must be taken in choosing the constant in order that the angles change continuously with the eigenvalues.

The construction goes as follows. Because V and $\tilde{V}$ are closed convex sets not containing the origin, there must be a ray R extending from the origin that does not intersect $V \cup \tilde{V}$. For any point (α,β) define $\theta(\alpha,\beta)$ to be the angle, measured clockwise, subtended by R and the ray from $(0,0)$ through (α,β) . For any eigenvalue λ_i of (1.1) let α_i and β_i be the Rayleigh components of λ_i , and set $\theta_i = \theta(\alpha_i,\beta_i)$. We shall assume that these eigenangles are ordered so that

$$\theta_1 \leq \theta_2 \leq \cdots \leq \theta_n\ .$$

The eigenangles $\tilde{\theta}_i$ for the perturbed problem are defined similarly.

We are now in a position to state our basic result.

<u>Theorem</u> 6.2. If $\varepsilon < \gamma(A,B)$, then

$$|\theta_i - \tilde{\theta}_i| \leq \sin^{-1}\frac{\varepsilon}{\gamma(A,B)} \qquad (i = 1,2,\ldots,n)\ . \tag{6.4}$$

There are several comments to be made about this theorem. First, although the definition of the eigenangles involves an arbitrary choice of origin in the form of the ray R , this arbitrariness disappears in the theorem, which is stated in terms of differences of the eigenangles.

Second, the results hold uniformly for all eigenangles regardless of multiplicity. In this respect the result is analogous to the classical theorem for the Hermitian eigenvalue problem.

Third, the theorem implies that

$$\chi(\lambda_i,\tilde{\lambda}_i) \leq \frac{\varepsilon}{\gamma(A,B)}\ .$$

However, because $|\theta[(\alpha,\beta)] - \theta[(-\alpha,-\beta)] = \pi$ while $\chi[(\alpha/\beta),(-\alpha/-\beta)]=0$, the inequality (6.4) is a stronger statement than (6.5).

Fourth, the theorem implies the usual result for the Hermitian eigenvalue problem, i.e. if $B = I$ and $F = 0$, then $|\lambda_i - \tilde{\lambda}_i| \leq \|E\|$. However, it cannot be obtained by replacing $\gamma(A,B)$ in (6.5) with $\gamma(A,I)$. Rather one must consider the bounds obtained for the problem $\tau Ax = \lambda(\tau)x$ as τ approaches zero. This underscores a difficulty with the bound (6.5); namely different scalings of A and B give essentially different bounds. Exactly how to scale A and B is an open question worthy of further investigation.

Finally, we note that for a particular eigenvalue the results are not asymptotically sharp. For example, the first order theory leads us to expect perturbations in a simple eigenvalue λ_i of order ε/γ_i where

$$\gamma_i = \sqrt{\alpha_i^2 + \beta_i^2} \geq \gamma(A,B) .$$

This is the price that must be paid for uniformity and freedom from considerations of multiplicity.

7. Notes and references.

The algebraic theory of the generalized eigenvalue problem goes back at least to Kronecker and Wierstrauss, who deveolped canonical forms for A and B under equivalence transformations. A treatment of these results may be found in Gantmacher [4].

It is very natural, both in theory and computational practice, to attempt to reduce the generalized eigenvalue problem to the ordinary one whose perturbation theory is well understood. Householder [5], Kato [6], and Wilkinson [12] give surveys of this theory, and the last named is the authoritative reference for computational methods. The approach is successful for many classes of problems, and the literature is studded with special results, far too many to cite here. The drawbacks of the approach are obvious to anyone who tries it. It is interesting to note that an ill-condition B causes not only theoretical problems but also computational problems associated with the formation of $B^{-1}A$ and the subsequent computation of its eigenvalues.

The results of §4-6 summarize work done by the author over several years [7,9,10]. The essence of the approach is a thoroughgoing symmetry in the treatment of A and B, so that the problem (1.1) and the reciprocal problem (3.2) are equivalent. The use of the chordal metric was suggested to me by W. Kahan, after I had developed the first order theory in terms of $\tan^{-1}\lambda$. The Gerschgorin theory is new; however, its proof and application are straightforward extensions of material appearing in the literature. In particular the technique for reducing the radii of Gerschgorin disks has been extensively employed by Wilkinson [12].

Theorem 6.1 is known as Calabi's theorem [1], and it has a fairly large literature of its own, for which see the forthcoming survey by Uhlig [11]. The proof based on the convexity of the field of values of a matrix is the author's; however, the idea of rotating the set V until it is in the upper half plane is due to Crawford [2]. In the case of real A and B it is natural to restrict x in the definition of $\gamma(A,B)$ to be real. This produces the same value of $\gamma(A,B)$, except, curiously, when $n = 2$, in which case the value can be nonzero while the true value is zero.

Crawford [2] was the first to realize that the value of $\gamma(A,B)$ plays a role in the perturbation theory, and he obtains bounds that are slightly weaker than the ones given here. The proof of Theorem 6.2 is like the proof of the corresponding theorem for the Hermitian eigenvalue problem. It is based on a min-max characterization of eigenangles,

$$\theta_i = \min_{\dim(X)=i} \max_{\substack{x \in X \\ x \neq 0}} \theta(x^H Ax,\ x^H Bx) ,$$

which also corresponds to a classical result.

The perturbation theory for eigenvectors is less well developed. Part of the problem is the necessity of working with subspaces when eigenvalues occur in clusters, a problem which arises in the ordinary eigenvalue problem (e.g. see [3] and [8]). The author has obtained some results in [7] and [10]; however more work on this problem is needed.

REFERENCES

1. E. Calabi (1964), Linear systems of real quadratic forms, Proc. Amer. Math. Soc., 15, pp. 844-846.

2. C. R. Crawford (1976), A stable generalized eigenvalue problem, SIAM J. Numer. Anal., 6, pp. 854-860.

3. C. Davis and W. M. Kahan (1970), The rotation of eigenvectors by a perturbation. III, SIAM J. Numer. Anal., 7, pp. 1-46.

4. F. R. Gantmacher (1960), The Theory of Matrices, Chelsea, New York.

5. A. S. Householder (1968), The Theory of Matrices in Numerical Analysis, Blaisdell, New York.

6. T. Kato (1966), Perturbation Theory for Linear Operators, Springer, New York.

7. G. W. Stewart (1972), On the sensitivity of the eigenvalue problem $Ax = \lambda Bx$, SIAM J. Numer. Anal., 9, pp. 669-686.

8. ______________ (1973), Error and perturbation bounds for subspaces associated with certain eigenvalue problems, SIAM Rev., 15, pp. 772-764.

9. ______________ (1975), Gerschgorin theory for the generalized eigenvalue problem $Ax = \lambda Bx$, Math. Comp., 29, pp. 600-606.

10. ______________ (1977), Perturbation bounds for the definite generalized eigenvalue problem, University of Maryland TR-591, to appear in Lin. Alg. Appl.

11. F. Uhlig (1978), A recurring theorem about pairs of quadratic forms and extensions, Institut für Geometrie und Praktische Mathematic, Aachen, W. Germany, to appear in Lin. Alg. Appl.

12. J. H. Wilkinson (1965), The Algebraic Eigenvalue Problem, Clarendon Press, Oxford.

The author was partially supported by the Office of Naval Research under Contract No. N00014-76-C-0391.

Department of Computer Science
University of Maryland
College Park, Maryland 20742

Some Remarks on Good, Simple, and Optimal Quadrature Formulas

H. F. Weinberger

1. INTRODUCTION.

There are two properties which one would like a numerical quadrature formula

$$N^{-1} \sum_{j=0}^{N} a_j u(j/N) \tag{1.1}$$

for approximating the integral $\int_0^1 u dx$ to possess. One is that it be accurate in the sense that the error is small.

The other desirable property is that the formula be computationally simple. In practice, this means that only a small number, independent of N, of the weights a_j should differ from 1, in order to avoid problems of excessive storage, running time, and roundoff errors.

It is the purpose of this paper to show that the two desired properties are not as incompatible as one might fear. In fact we shall show how to construct, for any given order, a whole family of quadrature formulas of that order which are both simple in the above sense and whose error bound, while not quite minimal, is asymptotically equal to the minimal error bound when N is large. We shall say that such a quadrature formula is good.

The simplest of our good rules of order K is the Gregory rule (3.1).

Our concept of an error bound is that of A. Sard [7]. The formula (1.1) is said to be K^{th} order accurate if it gives the exact integral for all polynomials of degree $K-1$. The error bound A corresponding to the formula (1.1) is then the smallest constant for which the inequality

ISBN 0-12-208360-1

$$\left| \int_0^1 u dx - N^{-1} \sum_{j=0}^{N} a_j u(j/N) \right| \leq A \{ \int_0^1 u^{[K]^2} dx \}^{1/2}$$

is valid. Sard defined the quadrature formula (1.1) to be optimal if its error bound A is as small as possible.

Sard showed that the problem of finding the optimal quadrature formula can be reduced to solving N linear equations in N unknowns, and solved some special cases. In particular, he observed that for $K = 1$ the trapezoidal rule is optimal, but for higher K the optimal schemes are by no means simple.

L.F. Meyers and A. Sard [6] obtained formulas for the optimal coefficients a_j when $K = 2$ as functions of N, and also tabulated the optimal coefficients a_j for $K = 3$ and 4 and various small values of N. They observed that the optimal coefficients converged as $N \to \infty$ when $K = 2$ and conjectured that the same is true for higher K. This conjecture was proved by I.J. Schoenberg [8]. Formulas for these limiting values were found and tabulated by I.J. Schoenberg and S.D. Silliman [10].

M. Golomb and the author [4] considered a class of problems of which the following is a special case. Given the values $u(0), u(1/N), \ldots, u(1)$ of a function u and a bound for the integral $\int_0^1 u^{[K]^2} dx$, what are the possible values that the linear functional $\int_0^1 u dx$ may have? The answer is that the set of possible values is an interval whose midpoint is the result of applying Sard's optimal quadrature scheme to the given values $u(j/N)$, and whose length is twice the error bound for this optimal scheme times the square root of the bound for $\int_0^1 u^{[K]^2} dx$.

The author showed [12] that as long as one works with the values of u at equally spaced points, the optimal scheme can be found by solving two sets of $K - 1$ equations in $K - 1$ unknowns instead of N equations in N unknowns. This makes the problem much more tractable.

In Section 2 we shall give a slightly different version of this idea, which enables us not only to obtain the optimal scheme for any K and all N in a rather explicit form, but also to obtain the first term of a large-N asymptotic formula for the optimal error bound. This term is, of course, a constant times N^{-K}.

In Section 3 we use the same formulation to show how to construct a large class of simple schemes whose error bounds have asymptotic expressions with the same first term as that of the optimal scheme. That is, they are both simple and good.

The simplest such rule is the Gregory rule, in which all the $a_j = 1$ except for K of them at each end. The asymptotic error in the Gregory rule of order four is shown to be about $1/8$ the corresponding term in Simpson's rule. Thus Gregory's rule, which is simpler to program than Simpson's rule, also gives better results, at least when N is large.

In Section 4 we extend these results to the approximation of an integral over a subinterval of the interval $[0,1]$ where the data are given. We again find quadrature rules which are both simple and good.

In Section 5 we consider the somewhat more general problem of tabulating the indefinite integral $Su \equiv \int_0^x u(t)dt$ in such a way that when a prescribed interpolation scheme is applied to the tabulated values, we obtain a good approximation in norm to this indefinite integral. This problem is in a class of problems that was treated but not solved completely in [13]. (The complete solution has been found by C. Davis, W.M. Kahan, and the author [1]). We shall show that in a special case of our problem we can combine the results of [13] and of Section 4 to find a simple rather good scheme for tabulating the integral. By rather good, we mean that the leading term of the asymptotic series for the error bound is at most $\sqrt{2}$ times the corresponding term for the optimal scheme.

Throughout this work we define the error bounds in terms of quadratic norms, because they are easier to deal with. One is more likely to be given a bound for the maximum norm rather than the L_2 norm of $u^{[K]}$.

It would be useful to see whether there are quadrature rules which are both simple and good in the sense of such norms.

2. OPTIMAL QUADRATURE FORMULAS.

We seek to find the function $\psi(x)$ of the form

$$\psi = \frac{(-1)^h}{K!}\{x^K - \frac{K}{N}\sum_{j=0}^{N-1} a_j(x-\frac{j}{N})_+^{K-1}\} \tag{2.1}$$

which satisfies the boundary conditions

$$\psi(1) = \psi'(1) = \dots = \psi^{[K-2]}(1) = 0 \quad . \tag{2.2}$$

and makes $\int_0^1 \psi^2 dx$ as small as possible. An integration by parts shows that for any ψ with the above properties and for any function u in $C^K[0,1]$

$$\int_0^1 u dx - N^{-1} \sum_{j=0}^{N} a_j u(\tfrac{j}{n}) = \int_0^1 \psi\, u^{[K]} dx \quad , \tag{2.3}$$

provided we define

$$a_N = N - \sum_{j=0}^{N-1} a_j \quad .$$

Schwarz's inequality shows that

$$\left| \int_0^1 u dx - N^{-1} \sum_{j=0}^{N} a_j u(\tfrac{j}{N}) \right| \le \{\int_0^1 \psi^2 dx\}^{1/2} \{\int_0^1 u^{[K]^2} dx\}^{1/2} \tag{2.4}$$

with equality when $u^{[K]} = \psi$. This error bound is to be made small, and the coefficients a_j which correspond to the ψ with minimal L_2 norm give the optimal quadrature formula in the sense of Sard [7].

In order to simplify our computations, we first define a function $\varphi(y)$ on the interval $[-1,1]$ as the solution of the problem

$$\begin{aligned} &\varphi^{[K]}(y) = (-1)^K && \text{for} \quad -1 < y < 1 \\ &\varphi^{[\ell]}(-1) = \varphi^{[\ell]}(1) && \text{for} \quad \ell = 0,\ldots,K-2 \\ &\int_{-1}^{1} \varphi\, dy = 0 \quad . \end{aligned} \tag{2.5}$$

It is easily seen that this problem has a unique solution, and that φ is even when K is even and odd when K is odd.

The function $(2N)^{-K}\varphi(2Nx - 2[Nx] - 1)$, where $[y]$ denotes the largest integer below y , is periodic of period $1/N$. The function

$$\psi = (2N)^{-K}\varphi(2Nx - 2[Nx]-1) - N^{-1} \sum_{j=-K+1}^{N-1} c_j \Delta^K (x - \tfrac{j}{N})_+^{K-1}, \tag{2.6}$$

is easily seen to be of form (2.1), provided $\psi(0) = \psi'(0) = \ldots = \psi^{[K-2]}(0) = 0$. Here Δ^K is the Kth forward difference operator

$$\Delta^K d_j = \sum_{m=0}^{K} d_{j+m} \binom{K}{m} (-1)^m \quad .$$

The function $\Delta^K (x - \frac{j}{N})_+^{K-1}$ clearly vanishes outside the interval $(j/N, (j+K)/N)$. (In the terminology of Schoenberg, it is a B-spline.)

Consequently, the end conditions on ψ are equivalent to the system of $K-1$ linear equations

$$\sum_{j=-K+1}^{-1} c_j \Delta^K(-j)_+^{K-\ell-1} = 2^{\ell-K}\frac{(K-\ell-1)!}{(K-1)!}\varphi^{[\ell]}(-1)\ , \ \ell=0,1,\ldots,K-2 \quad (2.7)$$

which determine $c_{-K+1},\ldots,c_{-1}$. Note that these numbers do not depend upon N . The symmetry of the coefficients shows that

$$c_{-K-j} = (-1)^K c_j \quad \text{for} \quad j=-K+1,\ldots,-1 \ . \quad (2.8)$$

Thus ψ , as given by (2.6), is of the form (2.1) if and only if $c_{-K+1},\ldots,c_{-1}$ are determined by (2.7). The end conditions (2.2) lead to the system

$$\sum_{j=N-K+1}^{N-1} c_j \Delta^K(N-j)_+^{K-\ell-1} = 2^{\ell-K}\frac{(K-\ell-1)!}{(K-1)!}\varphi^{[\ell]}(1),\ \ell=0,\ldots,K-2 \ .$$

Because of the conditions $\varphi^{[\ell]}(-1) = \varphi^{[\ell]}(1)$, the change $j' = j-N$ of the summation variable and (2.8) show that

$$c_j = c_{j-N} = (-1)^K c_{N-K-j} \quad \text{for} \quad j=N-K+1,\ldots,N-1 \ . \quad (2.9)$$

Thus a ψ of the form (2.6) gives a quadrature formula as in (2.3) if and only if the c_j for $j \le 0$ and $j \ge N-K+1$ are determined by (2.7) and (2.9). Our problem of minimization then reduces to choosing $c_0,\ldots,c_{N-K}$ to minimize the L_2 norm of ψ .

This minimum is characterized by setting the derivative with respect to each variable c_j equal to zero. That is, $c_0,\ldots,c_{N-K}$ are to be determined from the equations

$$\int_0^1 \psi\, \Delta^K\left(x-\frac{j}{N}\right)_+^{K-1} dx = 0 \quad \text{for} \quad j=0,\ldots,N-K \ .$$

It is easily seen that because of the last condition on φ and because $\varphi(2Nx-2[Nx]-1)$ is periodic, this function is orthogonal to the B-splines $\Delta^K\left(x-\frac{j}{n}\right)_+^{K-1}$ for $j=0,\ldots,N-K$. Consequently, our equations reduce to

$$\sum_{j=-K+1}^{N-1} c_j \int_0^1 \Delta^K\left(x-\frac{j}{N}\right)_+^{K-1} \Delta^K\left(x-\frac{\ell}{N}\right)_+^{K-1} dx = 0,\ \ell=0,\ldots,N-K \ .$$

Integration by parts shows that

$$\int_0^1 \Delta^K(x-\frac{j}{N})_+^{K-1} \Delta^K(x-\frac{\ell}{N})_+^{K-1} dx$$

$$= \frac{(K-1)!^2}{(2K-1)!} (-1)^K \Delta_j^K \Delta_\ell^K (\frac{j-\ell}{N})_+^{2K-1} \tag{2.10}$$

$$= \frac{(K-1)!^2}{(2K-1)!} (-1)^K N^{1-2K} \Delta_j^{2K} (j-\ell-K)_+^{2K-1} .$$

Hence our equations become

$$\sum_{j=-K+1}^{N-1} c_j \Delta_j^{2K} (j-\ell-K)_+^{2K-1} = 0 \qquad \text{for } \ell=0,\ldots,N-K \tag{2.11}$$

with c_j prescribed for $j \leq -1$ and $j \geq N-K+1$. Our problem is thus reduced to solving a boundary value problem for a finite difference equation with constant coefficients.

If z is any zero of the polynomial

$$Q_{2K-1}(z) = \sum_{j=0}^{2K-2} \Delta^{2K}(j-1)_+^{K-1} z^j ,$$

then $c_j = z^j$ is a solution of the equation (2.11). The polynomial Q_{2K-1} is the Eulerian (or Euler-Frobenius) polynomial of degree $2K-2$ [2,9]. As I.J. Schoenberg [9] has pointed out, it has been rediscovered many times. (See, e.g. [12].) It is known [3,9] that its zeros are real, distinct, and negative. Because of the symmetry of the coefficients, if z is a zero, so is $1/z$. We denote the zeros of Q_{2K-2} by $z_1 < z_2 < \ldots < z_{K-1} < z_{K-1}^{-1} < \ldots < z_1^{-1}$. Then the general solution of the equation (2.11) is

$$c_j = \sum_{\nu=1}^{K-1} (A_\nu z_\nu^j + B_\nu z_\nu^{-j}) . \tag{2.12}$$

The coefficients A_ν and B_ν are determined from the $K-1$ values $c_{-K+1},\ldots,c_{-1}$, at the left end and the $K-1$ values $c_{N-K+1},\ldots,c_{N-1}$ at the right end which have been determined from (2.7) and (2.9) .

Because Q_{2K-1} has $K-1$ zeros of absolute value above 1 and $K-1$ of absolute value below 1 , the contribution to the solution from each and decays exponentially as the distance from that end becomes large.

Because of the symmetry (2.9), we find that $c_{N-K-j} = (-1)^K c_j$ for all j . Consequently, $B_\nu = (-1)^K z_\nu^{N-K} A_\nu$, and we can find the A_ν by solving (2.12) with $j = -K+1,\ldots,-1$.

By applying summation by parts to (2.6), we can write ψ in the form

$$\psi = (2N)^{-K}\varphi(2Nx - 2[Nx]-1) - (-1)^K N^{-1} \sum_{j=K+1}^{N-1} (x - \frac{j}{N})_+^{K-1} \Delta^K c_{j-K}, \qquad (2.12)$$

where we define $c_j = 0$ for $j \leq -K$. From this it is easy to find the coefficients a_j of the corresponding quadrature rule. Integration by parts shows that

$$\begin{aligned} a_0 &= \frac{1}{2} + (-1)^K(K-1)!\,\Delta^{K-1}c_{j-K+1}\Big|_{j=0} \quad \text{for } j=0,N, \\ a_j &= 1+(-1)^K(K-1)!\,\Delta^K c_{j-K} \quad 1 \leq j \leq N-1 . \end{aligned} \qquad (2.13)$$

(We have used the fact that the function $(2N)^{-K}\varphi(2Nx-2[Nx]-1)$ gives the Euler-Maclaurin formula, which is the trapezoidal rule plus a linear combination of the derivatives of u at the end points.)

We shall give two examples.

When $K=2$ we obtain the Eulerian polynomial $Q_3(z) = z^2 + 4z + 1$, whose zeros are $z_1 = -2 - \sqrt{3}$ and $z_2^{-1} = -2 + \sqrt{3}$. We find that $\varphi(y) = \frac{1}{2} y^2 - \frac{1}{6}$ and that $c_{-1} = c_{N-1} = 1/12$. We immediately see that

$$c_j = \frac{z_1^{-j-1} + z_1^{j+1-N}}{12(1+z_1^{-N})} .$$

Therefore

$$a_j = \begin{cases} \dfrac{5}{12} + \dfrac{z_1^{-1}+z_1^{1-N}}{12(1+z_1^{-N})}, & j = 0,N \\[2ex] 1 - \dfrac{z_1^{-j}+z_1^{j-N}}{2(1+z_1^{-N})} & 1 \leq j \leq N-1 \end{cases} . \qquad (2.14)$$

This formula was obtained in another way by L.F. Meyers and A. Sard [6].

For $K = 4$ we find that

$$Q_7(z) = z^6 + 120z^5 + 1191z^4 + 2416z^3 + 1191z^2 + 120z + 1 .$$

Its zeros

$$z_1 = -109.3$$

$$z_2 = -8.16$$

$$z_3 = -1.87$$

have been computed to a high degree of accuracy by I.J. Schoenberg and S.D. Silliman [10].

We have

$$\varphi(y) = \frac{1}{24}\{(y^2-1)^2 - \frac{8}{15}\} ,$$

$$c_{-3} = c_{-1} = 19/4320 \text{ , and } c_{-2} = -11/4320 \text{ .}$$

We find that

$$c_j = \frac{\begin{vmatrix} z_1^{-j}+z_1^{j+4-N} & z_2^{-j}+z_2^{j+4-N} & z_3^{-j}+z_3^{j+4-N} & 0 \\ z_1+z_1^{3-N} & z_2+z_2^{3-N} & z_3+z_3^{3-N} & 19 \\ z_1^2+z_1^{2-N} & z_2^2+z_2^{2-N} & z_3^2+z_3^{2-N} & -11 \\ z_1^3+z_1^{1-N} & z_2^3+z_2^{1-N} & z_3^3+z_3^{1-N} & 19 \end{vmatrix}}{4320\begin{vmatrix} z_1+z_1^{3-N} & z_2+z_2^{3-N} & z_3+z_3^{3-N} \\ z_1^2+z_1^{2-N} & z_2^2+z_2^{2-N} & z_3^2+z_3^{2-N} \\ z_1^3+z_1^{1-N} & z_2^3+z_2^{1-N} & z_3^3+z_3^{1-N} \end{vmatrix}}$$

Therefore, for $K = 4$

$$a_0 = a_N = \frac{1}{2} + \frac{\begin{vmatrix} (1-z_1)^3(1-z_1^{1-N}) & (1-z_2)^3(1-z_2^{1-N}) & (1-z_3)^3(1-z_3^{1-N}) & 0 \\ z_1+z_1^{3-N} & z_2+z_2^{3-N} & z_3+z_3^{3-N} & 19 \\ z_1^2+z_1^{2-N} & z_2^2+z_2^{2-N} & z_3^2+z_3^{2-N} & -11 \\ z_1^3+z_1^{1-N} & z_2^3+z_2^{1-N} & z_3^3+z_3^{1-N} & 19 \end{vmatrix}}{720\begin{vmatrix} z_1+z_1^{3-N} & z_2+z_2^{3-N} & z_3+z_3^{3-N} \\ z_1^2+z_1^{2-N} & z_2^2+z_2^{2-N} & z_3^2+z_3^{2-N} \\ z_1^3+z_1^{1-N} & z_2^3+z_2^{1-N} & z_3^3+z_3^{1-N} \end{vmatrix}},$$

(2.15)

$$a_j = 1 + \frac{\begin{vmatrix} (1-z_1)^4(z_1^{-j}+z_1^{j-N}) & (1-z_2)^4(z_2^{-j}+z_2^{j-N}) & (1-z_3)^4(z_3^{-j}+z_3^{j-N}) & 0 \\ z_1+z_1^{3-N} & z_2+z_2^{3-N} & z_3+z_3^{3-N} & 19 \\ z_1^2+z_1^{2-N} & z_2^3+z_2^{2-N} & z_3^2+z_3^{2-N} & -11 \\ z_1^3+z_1^{1-N} & z_2^3+z_2^{1-N} & z_3^3+z_3^{1-N} & 19 \end{vmatrix}}{720\begin{vmatrix} z_1+z_1^{3-N} & z_2+z_2^{3-N} & z_3+z_3^{3-N} \\ z_1^2+z_1^{2-N} & z_2^2+z_2^{2-N} & z_3^2+z_3^{2-N} \\ z_1^3+z_1^{1-N} & z_2^3+z_2^{1-N} & z_3^3+z_3^{1-N} \end{vmatrix}}$$

for $1 \leq j \leq N-1$.

It was noticed by Meyers and Sard [6] that for the case $K=2$ the coefficients a_j given by (2.14) have a limit as $N \to \infty$, and it was conjectured by Meyers and Sard and proved by I.J. Schoenberg [8] that the same is true for all K . The formulas (2.14) and (2.15) not only make the convergence clear, but say something about the rate of convergence. We see from (2.14) that if H is so large that $|z_1|^{-N/2} = (2+\sqrt{3})^{-N/2}$ is negligible relative to 1 , we can neglect the second term in the denominator and, if $j \leq \frac{1}{2}N$, the first term in the numerator. That is,

$$a_j \cong \begin{cases} \frac{1}{12}(5 - z_1^{-1}) & j=0 \\ 1 - \frac{1}{2}\, z_1^{-j} & 1 \leq j \leq \frac{1}{2}N \end{cases}$$

Similarly, we see from (2.16) that if $K = 4$ and N is so large that we may neglect $|z_3|^{-N}$ relative to 1, then $a_j - 1$ is a linear combination independent of N of z_1^{-j}, z_2^{-j}, and z_3^{-j} for $1 \leq j \leq N-1$.

These limiting values are precisely what one obtains for the optimal approximation to $\int_0^\infty u\,dx$. The corresponding formulas were discovered and called semicardinal spline quadrature formulas by I.J. Schoenberg [8]. They have been tabulated by I.J. Schoenberg and S.D. Silliman [10]. We list approximate values of the first few of these limiting coefficients for $K = 2$ and $K = 4$ here:

TABLE 1

<u>K=2</u>: $a_0 = .394$, $a_1 = 1.133$, $a_2 = .964$, $a_3 = 1.0096$, $a_4 = .9974$, $a_5 = 1.00069$, $a_6 = .999815$, $a_7 = 1.000050$

<u>K=4</u>: $a_0 = .332$, $a_1 = 1.321$, $a_2 = .739$, $a_3 = 1.170$, $a_4 = .905$, $a_5 = 1.051$, $a_6 = .973$, $a_7 = 1.0147$, $a_8 = .9921$, $a_9 = 1.0042$, $a_{10} = .9977$, $a_{11} = 1.0012$, $a_{12} = .99935$.

We note that for each K these limiting coefficients a_j converge to 1 exponentially, roughly like $|z_{K-1}|^{-j}$, as j increases, provided j remains below $N/2$. Because z_{K-1} is negative, the coefficients oscillate about 1. This convergence was noted for $K=2$ and conjectured for other K by Meyers and Sard [6] and proved by Schoenberg [8].

Because $|z_{K-1}|$ decreases as K increases, the convergence becomes slower and slower with increasing K.

We now recall that, according to (2.4), the L_2 norm of the optimal ψ gives the error bound for the corresponding quadrature formula.

If ψ is given by (2.6),

$$\int \psi^2 dx = \int_0^1 (2N)^{-2K} \varphi(2Nx - 2[Nx]-1)^2 dx$$

$$-2N^{-1} \sum_{j=K+1}^{N-1} c_j (2N)^{-K} \int_0^1 \varphi(2Nx-2[Nx]-1) \Delta^K (x - \tfrac{j}{N})_+^{K-1}\, dx$$

$$+N^{-2} \sum_{j,\ell=K+1}^{N-1} c_j\, c_\ell \int_0^1 \Delta^K (x - \tfrac{j}{N})_+^{K-1}\, \Delta^K (x - \tfrac{\ell}{N})_+^{K-1} dx \quad .$$

A splitting of the interval into N subintervals and a suitable change of variables show that

$$\int_0^1 \varphi(2Nx - 2[Nx]-1)^2 dx = \frac{1}{2}\int_{-1}^1 \varphi(y)^2 dy \quad .$$

We recall that $\varphi(2Nx-2[Nx]-1)$ is orthogonal to the B-splines $\Delta^K(x-\frac{j}{N})_+^{K-1}$ for $0 \le j \le N-K$, that the optimal coefficients c_j satisfy (2.11) for $0 \le \ell \le N-K$, and that $\Delta^K(x-\frac{j}{N})_+^{K-1}\Delta^K(x-\frac{\ell}{N})_+^{K-1} \equiv 0$ for $|j-\ell| \ge K$. Therefore

$$\int \psi^2 dx = N^{-2K}2^{-2K-1}\int_{-1}^1 \varphi(y)^2 dy$$

$$-4(2N)^{-K-1}\sum_{j=-K+1}^{-1} c_j\int_0^1 \varphi(2Nx-2[Nx]-1)\,\Delta^K(x-\frac{j}{N})_+^{K-1}dx \qquad (2.16)$$

$$+2N^{-2}\sum_{j=-K+1}^{K-2}\;\sum_{\ell=-K+1}^{-1} c_j c_\ell\int_0^1 \Delta^K(x-\frac{j}{N})_+^{K-1}\Delta^K(x-\frac{\ell}{N})_+^{K-1}dx \quad .$$

We have used the symmetry $c_j = (-1)^K c_{N-K-j}$. (It should be noted that the formula (2.10) is not valid if neither ℓ nor j is on the interval $[0,N-K]$.)

We now observe that φ is independent of N and that $N^{K-1}\Delta^K(x-\frac{j}{N})_+^{K-1}$ is bounded and vanishes outside an interval of length K/N. Moreover the number of terms in the sums is independent of N. It is easily seen that the optimal c_j are bounded uniformly in N. We conclude from these facts that

$$\int_0^1 \psi^2 dx = N^{-2K}2^{-2K-1}\int_{-1}^1 \varphi^2 dy + O(N^{-2K-1}) \;,$$

so that the optimal error bound has the form

$$\{\int_0^1 \psi^2 dx\}^{1/2} = 2^{-K-1/2}\{\int_{-1}^1 \varphi^2 dy\}^{1/2} N^{-K}(1+O(N^{-1})) \;. \qquad (2.17)$$

For example, for $K=2$ we find that

$$\int \psi^2 dx = \frac{1}{720\,N^4} + \frac{1}{216\,N^5}\{1 - \frac{2-\sqrt{3}}{2}\,\frac{1+(-2-\sqrt{3})^{2-N}}{1-(-2-\sqrt{3})^{-N}}\} \;. \qquad (2.18)$$

The error bound for the trapezoidal rule is $(120)^{-1/2}N^{-2}$ times the L_2 norm of u''. Thus for large N we gain a factor of $6^{1/2}$ by using the optimal quadrature rule.

For $K=4$ we find that

$$\{\int \psi^2 dx\}^{1/2} = \frac{1}{240\sqrt{21}\,N^4}\,(1+O(\tfrac{1}{N})) \quad .$$

The error term for Simpson's rule is $1/36\sqrt{14}\ N^4$. Thus for large N we gain a factor of $20/\sqrt{6} \cong 8$ by using the optimal quadrature rule.

3. GOOD SIMPLE QUADRATURE FORMULAS.

The preceding section shows how the weights in the optimal quadrature formula behave when the number of mesh points is large. The salient feature is that a_j is very close to 1 when j is far from 0 and N.

From the computational point of view, the optimal quadrature scheme is not very practical, because the optimal weights must either be stored, which creates problems of input and of storage, or they must be computed each time, which slows down the quadrature considerably. In any case the number of multiplications is at least of the order N, which increases both running time and roundoff error.

These drawbacks are hardly justified by the gain of a factor of $6^{1/2}$ in the second order rule or about 8 in the fourth order rule. A much more reasonable idea is to replace the optimal quadrature formula by one which is good in the sense that the error bound is small and which is, at the same time, computationally simple in the sense that all but a few of the weights are equal to 1.

We have seen in the preceding section that the coefficients c_j are near zero when j is far from both zero and $N\text{-}K$. This suggests the idea of requiring all but a fixed number of c_j with $0 \leq j \leq N\text{-}K$ in the formula (2.6) to be zero. We can then choose the remaining free c_j to minimize the L_2 norm of ψ.

Accordingly, we choose an integer M such that $K \leq M \leq N\text{-}K+2$, and set

$$c_j = 0 \quad \text{for} \quad M\text{-}K \leq j \leq N\text{-}M$$

in (2.6). Of course, c_j with $j \leq -1$ and $j \geq N\text{-}K+1$ must be determined from (2.7) and (2.9), as before. If we minimize the integral of ψ^2 with respect to the choice of $c_0, \ldots, c_{M-K-1}$, we obtain the equations (2.11) for $\ell = 0, \ldots, M\text{-}K\text{-}1$. In addition we are given $c_{-K+1}, \ldots, c_{-1}$, and we know that $c_{M-K} = c_{M-K+1} = \ldots = c_{M-1} = 0$. Thus we have a boundary value problem for our homogeneous difference equation, which can be solved as in the preceding section. We find that c_j depends on K and M, but not on N.

The coefficients c_j for $N\text{-}M+1 \leq j \leq N\text{-}K$ are found in the same way. We again find that $c_j = (-1)^K c_{N-K-j}$, so that it is not necessary to carry out this second solution process.

Once the c_j are found, we obtain the quadrature coefficients a_j from the formula (2.13). We see that

$$a_j = 1 \quad \text{for} \quad M \leq j \leq N-M \ .$$

Thus only M coefficients at each end differ from 1, so that this quadrature formula is simple when M is small.

Because the free c_j satisfy the equation (2.11), we again find the identity (2.16) for the L_2 norm of ψ. Because the coefficients c_j are independent of N, we derive the asymptotic form (2.17) as before. Thus when N is large, the error bound is essentially independent of the choice of M.

One might as well make the simplest choice $M=K$. That is, one sets $c_j = 0$ for $j=0,\ldots,N-K$, so that no minimization has to be done at all. This choice gives the Gregory quadrature rule with differences higher than $K-1$ deleted. That is, (see [5, p. 526] or [11, p. 64])

$$\int_0^1 u\,dx \cong N^{-1}\Big\{ \sum_{j=0}^{N} u(\tfrac{j}{N}) - \sum_{\ell=0}^{K-1} b_{\ell+1}\{ \Delta^{\ell} u(\tfrac{j}{N})\Big|_{j=0} + (-1)^{\ell}\nabla^{\ell} u(\tfrac{j}{N})\Big|_{j=N}\Big\} \qquad (3.1)$$

where b_m is the mth Bernoulli number of the second kind [5, p. 266] and ∇^{ℓ} is the ℓth backward difference.

When $K=2$ the choice $M=2$ gives

$$a_0 = a_N = \frac{5}{12} = .4166 \ldots$$

$$a_1 = a_{N-1} = \frac{13}{12} = 1.0833 \ldots$$

$$a_j = 1 \quad \text{for} \quad 2 \leq j \leq N-2 \ .$$

Note the oscillation below $\frac{1}{2}$ and above 1 of the end coefficients, which is reminiscent of the behavior of the optimal coefficients in Table 1. For the error bound we find

$$\int_0^1 \psi^2 dx = \frac{1}{720\,N^4} + \frac{1}{216\,N^5} \ .$$

Thus the effect of minimizing $\int \psi^2 dx$ with respect to the c_j instead of equating the c_j to zero is to subtract the second term in the braces in (2.18) from 1. This is clearly not a large effect.

It is obvious that the minimum value of the integral of ψ^2 is non-increasing in M, so that the bounds corresponding to various values of M are always between the optimum bound and the bound corresponding to $M=K$.

For $K = M = 4$ we find the coefficients

$$a_0 = a_N = \frac{251}{720} = .348611 \ldots$$

$$a_1 = a_{N-1} = \frac{299}{240} = 1.245833 \ldots$$

$$a_2 = a_{N-2} = \frac{211}{240} = .879166 \ldots$$

$$a_3 = a_{N-3} = \frac{739}{720} = 1.026388 \ldots$$

$$a_j = 1 \quad \text{for} \quad 4 \leq j \leq N-4 .$$

Note that a_0 is close to $1/3$, which occurs in Simpson's rule, and that the coefficients display an oscillation which is damped somewhat more quickly that that of the corresponding coefficients in Table 1.

Table 2 presents some double precision computations on the Cyber 74 of the integrals from 0 to 1 of the two functions $u = x^4 - .2$ and $u = x \sin 10\pi x^2 \cos^3 10\pi x^2 \,|\cos 10\pi x^2|$. Both of these integrals have the value 0. The second function is only piecewise C^4, and $u^{iv} \cong 10^3 u''$. Evaluations by the second and fourth order optimal methods and by their Gregory approximations are given. For comparison, we also present the results of using the trapezoidal rule and Simpson's rule. Note that not much is lost by replacing the optimal rule by the corresponding Gregory rule.

$u = x^4 - .2$:

TABLE 2

N	K=2 OPT.	K=2 M=2	TRAP.	K=4 OPT.	K=4 M=4	SIMP.
10	2.8×10^{-4}	4.7×10^{-4}	3.3×10^{-3}	1.0×10^{-5}	9.0×10^{-6}	1.3×10^{-5}
100	2.9×10^{-7}	5.0×10^{-5}	3.3×10^{-5}	3.9×10^{-11}	9.0×10^{-11}	1.3×10^{-9}
1000	2.9×10^{-10}	5.0×10^{-10}	3.3×10^{-7}	3.9×10^{-16}	9.0×10^{-16}	1.3×10^{-13}

$u = x \sin 10\pi x^2 \cos^3 10\pi x^2 \,|\cos 10\pi x^2|$

N	K=2 OPT.	K=2 M=2	TRAP.	K=4 OPT.	K=4 M=4	SIMP.
10	3.1×10^{-3}	2.1×10^{-3}	-6.0×10^{-29}	3.4×10^{-3}	5.4×10^{-3}	-1.2×10^{-28}
100	2.5×10^{-4}	3.8×10^{-4}	5.9×10^{-4}	-2.8×10^{-4}	-1.2×10^{-5}	-8.2×10^{-4}
1000	9.1×10^{-9}	5.7×10^{-8}	5.2×10^{-6}	-1.8×10^{-9}	-6.0×10^{-9}	-1.9×10^{-8}

The errors in the optimal and good methods decrease more rapidly than N^{-K}, because the function $u^{[K]}$ in the bound (2.4) can be replaced by its projection perpendicular to those B-splines which are orthogonal to ψ. The additional rate of convergence depends upon u.

When $N = 10$, both the trapezoidal rule and Simpson's rule happen to give the exact value 0 for the second function. While (2.4) gives the best error bound if only a bound for the L_2 norm of $u^{[K]}$ is known, there is for any quadrature rule a space of codimension 1, namely the space of all functions such that $u^{[K]}$ is orthogonal to the corresponding ψ, for which the quadrature rule is exact.

4. INTEGRALS OVER SUBINTERVALS.

We now consider the more general problem of approximating $\int_a^b u\,dx$ in terms of given values of $u(j/N)$ $j = 0,\dots,N$ and a bound for $\int_0^1 u^{[K]^2}\,dx$ when $[a,b]$ is a subinterval of $[0,1]$. For the sake of exposition we shall take $a = 0$ and $0 < b < 1$.

We approach this problem with the tools of section 2. We again define the function $\varphi(y)$ as the solution of (2.5) and find the coefficients $c_{-K+1},\dots,c_{-1}$ from the equations (2.7). We now define a function

$$\psi_0(x) = \begin{cases} (2N)^{-K}\varphi(2Nx-2[Nx]-1) - N^{-1}\sum_{j=-K+1}^{-1} c_j\Delta^K(x-\frac{j}{N})_+^{K-1} & \text{for } x \le b \\ N^{-1}\sum_{i=2}^{K+1} d_i(\frac{[Nb]+i}{N}-x)_+^{K-1} & \text{for } x \ge b \end{cases} \tag{4.1}$$

where the coefficients d_i are determined in such a way that

$$\sum_{i=2}^{K+1} d_i(i-Nb+[Nb])^{K-1-\ell} = \frac{(K-1-\ell)!}{(K-1)!}\,2^{\ell-K}\,\varphi^{[\ell]}(2Nb-2[Nb]-1) \tag{4.2}$$

$$\text{for } \ell = 0,\dots,K-1 .$$

Of course, the d_i depend upon $b-[Nb]/N$. We suppose that $(K-1)/N < b < (N-2K-2)/N$. Then ψ_0 and its first $K-1$ derivatives are continuous at b, and ψ_0 and its first $K-2$ derivatives vanish at 0 and at 1. An integration by parts shows that there are coefficients $a_j^{(0)}$ such that

$$\int_0^b u\,dx - N^{-1}\sum_{j=0}^{N} a_j^{(0)}\,u(\tfrac{j}{N}) = \int_0^1 \psi_0\, u^{[K]}\,dx . \tag{4.3}$$

The most general formula of this kind can be obtained by setting

$$\psi = \psi_0 - N^{-1}\sum_{j=0}^{N-K} c_j \Delta^K (x - \tfrac{j}{N})_+^{K-1} . \tag{4.4}$$

The corresponding quadrature formula is

$$\int_0^b u dx - N^{-1}\sum_{j=0}^{N} a_j\, u(\tfrac{j}{N}) = \int_0^1 \psi\, u^{[K]} dx ,$$

and Schwarz's inequality gives an error bound with the constant

$$\{\int_0^1 \psi^2 dx\}^{1/2} .$$

The optimal formula in the sense of Sard is obtained by minimizing this L_2 norm of ψ. The minimizing equations are

$$N^{-1}\sum_{j=0}^{N-K} c_j \int_0^1 \Delta^K(x-\tfrac{j}{N})_+^{K-1}\Delta^K(x-\tfrac{\ell}{N})_+^{K-1}dx = \int_0^1 \psi_0 \Delta^K(x-\tfrac{\ell}{N})_+^{K-1}\, dx,$$

$$\ell = 0,\dots,N-K ,$$

which can be written in the form

$$\sum_{j=-K+1}^{N-1} c_j \Delta_j^{2K}(j-\ell-K)_+^{2K-1} = \frac{(2K-1)!(-N)^K}{(K-1)!^2}$$

$$\{2^{-K}\int_0^b \varphi(2Nx-2[Nx]-1)\Delta^K(x-\tfrac{\ell}{N})_+^{K-1}dx \tag{4.5}$$

$$+\sum_{i=2}^{K+1} d_i \int_b^{\frac{[Nb]+i}{N}} ([Nb]+i-Nx)^{K-1}\Delta^K(x-\tfrac{\ell}{N})_+^{K-1}dx\} ,$$

provided we define $c_j = 0$ for $j \geq N-K+1$.

We recall that, by construction, $\varphi(2Nx-2[Nx]-1)$ is orthogonal to $\Delta^K(x-\frac{\ell}{N})_+^{K-1}$ on the interval $(\ell/N, (\ell+K)/N)$. Consequently, the right-hand side of this equation is zero for $0 \leq \ell \leq N-K$ except for $Nb-K \leq \ell \leq Nb+K+1$.

Thus, (4.5) is an inhomogeneous difference equation with constant coefficients whose right-hand side is zero except near Nb. The boundary values c_j for $-K+1 \leq j \leq -1$ are given and $c_j = 0$ for $j = N-K+1,\dots,N-1$. The solution can be written as a contribution from the inhomogeneous boundary values at the left end plus a contribution from the inhomogeneity. For large N, the first of these decays exponentially in the distance from the left end. The second part decays exponentially in the distance $|j-Nb|$. We can then show that the optimal quadrature coefficients a_j have the property that, as $N \to \infty$, a_j converges to a sum of two coefficients r_j and s_j. Here r_j is again the coefficient in the semicardinal spline approximation to $\int_0^\infty u\,dx$ in terms of the values $u(j)$, $0 \le j < \infty$. The coefficient s_j is the coefficient in the cardinal spline approximation to $\int_{-\infty}^b u\,dx$ in terms of $u(j)$, $-\infty < j < \infty$, minus 1.

We now look at the corresponding error bound. Because the function ψ_0 is orthogonal to all the B-splines except for a number which is independent of N and because the c_j satisfy (4.5), we can show in the same manner as we derived (2.17) that for the optimal ψ

$$\{\int_0^1 \psi^2 dx\}^{1/2} = \{2^{-2K-1} b \int_{-1}^1 \varphi^2 dx\}^{1/2} N^{-K}(1+O(N^{-1})) . \tag{4.6}$$

Moreover, the same reasoning shows that also

$$\{\int_0^1 \psi_0^2\, dx\}^{1/2} = \{2^{-2K-1} b \int_{-1}^1 \varphi^2 dx\}^{1/2} N^{-K}(1+O(N^{-1})) . \tag{4.7}$$

Thus, for large values of N the quadrature rule which corresponds to the coefficients $a_j^{(0)}$ is good in the sense that its error is asymptotically the same as that of the optimal rule. We see from the definition (4.1) that $a_j^{(0)} = 1$ for $K \le j \le [Nb]-1$ and $a_j^{(0)} = 0$ for $[Nb]+K+2 \le j \le N$. Thus the quadrature rule is a simple one whose weights differ from those of the trapezoidal rule only near the end points of the integral.

We have chosen to put the altered weights on the right outside the interval $(0,b)$ in defining ψ_0. However, it is easy to choose K coefficients c_j in (4.4) which put the extra weights inside the interval or which put half of them inside and half outside. All the corresponding schemes will be both simple and good, and we can, if we wish, optimize with respect to the choices of some of the c_j to make the $O(N^{-1})$ term in (4.6) as small as possible.

5. TABULATING AN INDEFINITE INTEGRAL.

We now consider the following problem. We are again given the values $u(j/N)$, $j=0,\ldots,N$ of a function and a bound for the L_2 norm of $u^{[K]}$. We wish to compute approximate values of the indefinite integral

$$Su(y) \equiv \int_0^y u\,dx$$

for $y = 0, 1/N', 2/N', \ldots, 1$ in such a way that when a particular linear interpolation scheme is used to construct a function $v(y)$ out of these values, then v is a good approximation to Su.

The interpolation scheme is described by prescribing $N'+1$ functions h_α and setting

$$v(y) = \sum_{\alpha=0}^{N'} r_\alpha h_\alpha(y) \tag{5.1}$$

where r_α is the tabulated approximation to $\int_0^{\alpha/N'} u\,dx$. Each r_α is to depend linearly on the given values of u :

$$r_\alpha = \sum_{j=0}^{N} Q_{\alpha j}\, u(\tfrac{j}{N}) \qquad \alpha = 0,\dots,N' .$$

We choose a positive definite norm

$$\|u\|_B = \{\int_0^1 u^{[K]^2} dx + \Sigma\, u(\tfrac{j}{N})^2\}^{1/2} \tag{5.2}$$

where the sum extends over at least K points. Our problem is to choose the matrix $Q_{\alpha j}$ in such a way that the interpolated function v approximates $S u$ in the sense of the pseudonorm

$$\|w\|_{K'} = \{\int_0^1 |w^{[K']}|^2 dx\}^{1/2} .$$

More specifically, we wish to choose $Q_{\alpha j}$ in such a way that the operator norm

$$\sup_{\|u\|_B \leq 1} \|v - Su\|_{K'}$$

is as small as possible. (We must use a positive definite norm on B because v cannot equal Su for polynomials of degree $K-1$ unless the set of linear combinations of h_α happens to contain the polynomials of degree K which vanish at zero.)

It is clear that the solution of this problem depends not only on the integers K and K' but also upon the interpolation functions h_α . For example, one would certainly want to put different numbers (namely approximate Fourier sine coefficients) in the table if $h_\alpha = \sin \pi\alpha x$ than if the h_α correspond to piecewise linear interpolation.

Even if the h_α are designed to interpolate pointwise values, it is not clear that r_α should be chosen as the optimal approximation to the integral from 0 to α/N' . It is quite conceivable that r_α with a deliberate error at each point will give a better overall approximation to Su in the sense of the norm.

The present problem is a special case of one which was considered elsewhere [13] in another context. It can be formulated as follows: Let $\mathfrak{N}$ be the linear operator of discretization, which takes the Hilbert space B of functions with the norm (5.2) into the space of (N+1)-tuples:

$$\mathfrak{N}u \equiv \{u(\tfrac{j}{N}),\ j=0,\ldots,N\} \quad .$$

Let $\mathfrak{M}$ be the interpolation map (5.1) from the space of (N'+1)-tuples to the space Σ with norm $\|w\|_{K'}$. Our problem is to find an $(N'+1)\times(N+1)$-matrix Q such that the operator norm

$$\|\mathfrak{M}Q\mathfrak{N} - S\|$$

is minimized.

The space B splits naturally into the null space B_1 of $\mathfrak{N}$ and its orthogonal complement B_2 . The latter is a finite dimensional space of 2K-1-splines: The space Σ splits naturally into the space Σ_2 spanned by $h_0,\ldots,h_{N'}$ and its orthogonal complement Σ_1 . Then the operator S has the natural splitting

$$\begin{pmatrix} S_{11} & S_{12} \\ S_{21} & S_{22} \end{pmatrix}$$

where S_{ij} is the orthogonal projection onto Σ_i of the restriction of S to B_j . In the corresponding splitting of the approximation $\mathfrak{M}Q\mathfrak{N}$, we find that $(\mathfrak{M}Q\mathfrak{N})_{ij} = 0$ unless $i=j=2$, while a suitable choice of Q will make $(\mathfrak{M}Q\mathfrak{N})_{22}$ equal to any operator from B_2 to Σ_2 . Thus,

$$S - \mathfrak{M}Q\mathfrak{N} = \begin{pmatrix} S_{11} & S_{12} \\ S_{21} & C \end{pmatrix} \tag{5.3}$$

where C is an arbitrary operator from B_2 to Σ_2 . Our problem is thus equivalent to that of determining C to make the norm of the above operator as small as possible.

There are two obvious restrictions on how small this norm can be. Let P be the orthogonal projection onto the finite dimensional space B_2 and let Π be the orthogonal projection onto the span Σ_2 of the h_α . Then clearly $\mathfrak{N}(I-P)=0$ and $(I-\Pi)\mathfrak{M}=0$. Therefore $(S-\mathfrak{M}Q\mathfrak{N})(I-P) = S(I-P)$, and it follows that

$$\|S - \mathfrak{M}Q\mathfrak{N}\| \geq \|S(I-P)\| = \sup_{\mathfrak{N}u=0} \frac{\|Su\|_{K'}}{\|u\|_B} \quad .$$

Similarly, we find that

$$\|S - \mathcal{M} Q \mathcal{N}\| \geq \|(I - \Pi)S\| \ .$$

It was shown in [13] that these are the only restrictions on $\|S - \mathcal{M} Q \mathcal{N}\|$. That is, if we define

$$\mu = \max\{\|S(I-P)\|, \|(I-\Pi)S\|\} \ , \tag{5.4}$$

there exists an optimal matrix Q such that

$$\|S - \mathcal{M} Q \mathcal{N}\| = \mu \ .$$

Because the operator norm is essentially a maximum norm (it is the maximum of the square roots of the eigenvalues of $(S - \mathcal{M} Q \mathcal{N})^*(S - \mathcal{M} Q \mathcal{N})$), the optimal Q is usually not unique. This makes it difficult to determine Q explicitly.

An obvious guess for the operator C which minimizes the norm of the split operator in (5.3) is $C = 0$. However, it is easily seen that the norm of the family of symmetric 2×2 matrices

$$\begin{pmatrix} 1 & 1 \\ 1 & \alpha \end{pmatrix}$$

attains its minimum at $\alpha = -1$ and not at $\alpha = 0$, so that this conjecture is not correct. Nevertheless, it is shown in [13] that the choice $C=0$ (that is, $\mathcal{M} Q \mathcal{N} = S_{22}P = \Pi SP$) leads to a good error bound. Namely,

$$\|S - \Pi S P\| \leq \sqrt{2}\, \mu \ ,$$

where μ is the optimal error bound. Thus the scheme ΠSP is reasonably good, although it is not optimal or even asymptotically optimal.

A construction which gives an optimal matrix has been found by C. Davis, W.M. Kahan, and the author [1]. (A paper is in preparation, and we only state the result here.) An optimal approximation is given by the formula

$$\mathcal{M} Q \mathcal{N} u = [S_{22} + S_{21}(\mu^2 I - S_{11}^* S_{11})^{-1} S_{11}^* S_{12}]P \ , \tag{5.5}$$

where μ is defined by (5.4) , I is the identity on B_1 , and S_{11}^* is the adjoint of the operator S_{11} from B_1 to Σ_1 .

We consider the example where $K = 2$, $K' = 1$, and $\mathcal{M}$ is piecewise linear interpolation. That is, h_α is a piecewise linear function which is 1 at α/N' and 0 at β/N' when $\beta \neq \alpha$.

If $u \in B_1$, then $u(j/N) = 0$ for all j. One then sees from the corresponding Euler equation that

$$\int_{\frac{j}{N}}^{\frac{j+1}{N}} u^2 dx \leq \frac{1}{N^4\pi^4} \int_{\frac{j}{N}}^{\frac{j+1}{N}} u''^2 \, dx$$

with equality when $u = \sin N\pi x$. By adding these inequalities, we see that for u in B_1

$$\|Su\|_1 \leq \frac{1}{N^2\pi^2} \|u\|_B ,$$

so that

$$\|S(I-P)\| = (N\pi)^{-2} .$$

It is easily seen that, for $K'=1$ and piecewise linear interpolation, Πw is the piecewise linear function which is equal to w at all points of the form α/N' with α an integer. Therefore

$$((I-\Pi)Su)' = u(x) - N' \int_{\frac{\alpha}{N'}}^{\frac{\alpha+1}{N'}} u(y)dy \quad \text{for } \frac{\alpha}{N'} \leq x \leq \frac{\alpha+1}{N'} .$$

It follows from Poincaré's inequality [14, p. 45] that

$$\int_{\frac{\alpha}{N'}}^{\frac{\alpha+1}{N'}} ((I-\Pi)Su)'^2 dx \leq \frac{1}{N'^2\pi^2} \int_{\frac{\alpha}{N'}}^{\frac{\alpha+1}{N'}} u'^2 dx .$$

Hence

$$\|(I-\Pi)Su\|_1^2 \leq (N'\pi)^{-2} \int_0^1 u'^2 dx .$$

We define

$$\|u\|_B^2 = \int_0^1 u''^2 dx + u(0)^2 + u(1)^2 .$$

It is easily seen from the calculus of variations that

$$\int u'^2 \, dx \leq 2\|u\|_B^2 .$$

Consequently, we find that

$$\|(I-\Pi)S\| \leq 2^{1/2}(N'\pi)^{-1} .$$

When $u = x - \frac{1}{2}$, we find that $\|(I-\Pi)Su\|_1 = 12^{-1/2} N'^{-1}$ while $\|u\|_B = 2^{-1/2}$. Therefore,

$$\|(I-\Pi)S\| \geq 6^{-1/2} N'^{-1} .$$

Thus

$$.408 < N'\|I-\Pi)S\| < .451 .$$

The optimal error bound is the smaller of the two numbers $\|S(I-P)\|$, which depends on N, and $\|(I-\Pi)S\|$, which depends on N'. If the first of these numbers is much smaller than the second, we can reduce the number N of points where u is evaluated without increasing the optimal error. Similarly, if the second is smaller than the first, we can tabulate fewer values of the integral without increasing the error. An efficient scheme, then, makes $\|S(I-P)\|$ equal to $\|(I-\Pi)S\|$, which means that N' should be proportional to N^2. The reason is that we are using a second order quadrature scheme but only linear interpolation.

Because $\Pi w = w$ at the mesh points, the reasonably good scheme $\mathcal{M}Q\eta = \Pi SP$ in this case consists of letting r_α be the optimal quadrature of the preceding section for the integral of u from 0 to α/N'. In the interest of computational simplicity, one again replaces these optimal quadratures by good quadratures by using, for example, the function ψ_0 of the preceding section. (Some modifications have, of course, to be made when α/N' is near 0 or 1.)

The computation of the optimal approximation (5.5) leads to the difficult problem of inverting the operator $\mu^2 I - S_{11}^* S_{11}$. We shall not deal with this problem here.

BIBLIOGRAPHY

1. C. Davis, W.M. Kahan, and H.F. Weinberger. Norm-preserving dilations and their applications to optimal error bounds. Manuscript in preparation.

2. D. Foata and M.-P. Schützenberger. Théorie Géométrique des Polynômes Eulériens. Lecture Notes in Math #138, Springer, Berlin, 1970.

3. F.G. Frobenius. Über die Bernoullischen Zahlen und die Eulerschen Polynome. Sitzungsber. d. Preuss. Acad. d. Wiss., zu Berlin. Phys. - Math. Kl. (1910) pp. 809-847, also in Gesammelte Abhandlungen, Springer, Berlin, 1938 , v. III, pp. 440-478.

4. M. Golomb and H.F. Weinberger. Optimal approximation and error bounds. On Numerical Approximation, ed. R.E. Langer, U. of Wisconsin Press, 1959, pp. 117-190.

5. C.Jordan. Calculus of Finite Differences. Second edition, Chelsea, New York, 1950.

6. L.F. Meyers and A. Sard. Best approximate integration formulas. J. of Math. and Phys., 29 (1950), pp. 118-123.

7. A. Sard. Best approximate integration formulas; best approximation formulas. Amer. J. of Math., 71 (1949), pp. 80-91.

8. I.J. Schoenberg. Cardinal interpolation and spline functions VI. Semi-cardinal interpolation and spline formulae. J.d'Analyse Math. 27 (1974), pp. 159-204.

9. I.J. Schoenberg. Cardinal spline interpolation. C.B.M.S. Regional Conf. Series in Appl. Math. #12. S.I.A.M., Philadelphia, 1973.

10. I.J. Schoenberg and S.D. Silliman. On semicardinal quadrature formulae. Math. of Comp. 28 (1974), pp. 483-497.

11. J. Todd. A Survey of Numerical Analysis. McGraw-Hill, New York, 1962.

12. H.F. Weinberger. Optimal approximation for functions prescribed at equally spaced points. J. of Research of the Nat. Bureau of Standards B. Math. and Math. Phys. 65B (1961), pp. 99-104.

13. H.F. Weinberger. On optimal numerical solution of partial differential equations. S.I.A.M. J. on Numer. Anal. 9 (1972), pp. 182-198.

14. H.F. Weinberger. Variational Methods for Eigenvalue Approximation. C.B.M.S. Regional Conf. Series in Appl. Math. #15. S.I.A.M., Philadelphia, 1974.

This work was partly supported by the National Science Foundation through Grant MCS 76-06128 A01. The computations were partly supported by a grant from the University of Minnesota Computer Center.

School of Mathematics
University of Minnesota
206 Church Street
Minneapolis, MN 55455

Linear Differential Equations and Kronecker's Canonical Form

J. H. Wilkinson

1. INTRODUCTION

The solution of an <u>explicit</u> n × n system of linear differential equations of first order with constant coefficients

$$\dot{x} = Ax + f \tag{1.1}$$

can be expressed in terms of the Jordan canonical form (J.c.f.) of A. Indeed if

$$X^{-1}AX = J, \tag{1.2}$$

where X is non-singular, (1.1) is equivalent to the explicit system

$$\dot{y} = Jy + g, \tag{1.3}$$

where

$$x = Xy,\ g = X^{-1}f. \tag{1.4}$$

In terms of the new variables y the equations decouple. This may be adequately illustrated by considering the first block in J. If, for example, this is a block of order 3 involving λ_1 the corresponding equations are

ISBN 0-12-208360-1

$$\left.\begin{aligned} \dot{y}_1 &= \lambda_1 y_1 + y_2 + g_1 \\ \dot{y}_2 &= \lambda_1 y_2 + y_3 + g_2 \\ \dot{y}_3 &= \lambda_1 y_3 + g_3 \end{aligned}\right] \qquad (1.5)$$

which may be expressed in the form

$$\left.\begin{aligned} z_1 &= z_2 + h_1 \\ z_2 &= z_3 + h_2 \\ z_3 &= h_3 \end{aligned}\right] \qquad (1.6)$$

where

$$z_i = y_i \exp(-\lambda_1 t) \;, \; h_i = g_i \exp(-\lambda_1 t) \;. \qquad (1.7)$$

The general solution may be given in terms of quadratures. It involves three arbitrary constants and the initial values of the z_i may be prescribed arbitrarily; no relations between the h_i (and hence the f_i) are required for the solution to exist. The general solution may also be expressed in terms of exp(At) but this will not concern us here.

The relevance of the J.c.f. of A to the general solution of the explicit system is almost universally appreciated by numerical analysts, physicists and engineers. It is widely treated in the literature. However differential equations rarely arise in the explicit form (1.1). Much more commonly they arise in the <u>implicit</u> form

$$B\dot{x} = Ax + f \;. \qquad (1.8)$$

The relevant algebraic theory in this case is provided by the work of Weierstrass and Kronecker on the matrix pencil $A-\lambda B$. In my experience this work is not widely known to applied mathematicians in general and is rarely discussed by numerical analysts. Weierstrass and Kronecker developed their theory in terms of the elementary divisors of $A-\lambda B$. Whereas the tie-up between similarity theory and explicit differential systems is widely treated in the literature this is not true of the analogous tie-up between the Weierstrass-Kronecker theories and implicit systems, though to be sure, research workers in control theory do make wide use of this relationship.

In this paper we shall, in effect, redevelop a good part of the W-K theory using the differential system as our motivation. The methods used by W and K were not very promising algorithmically and indeed the very concept of elementary divisors does not lend itself to an algorithmic approach. It is well known that in spite of its importance in establishing the formal nature of the general solution of an explicit system, the J.c.f. is not as valuable in practice as might be expected because of its extreme sensitivity to perturbations in the matrix elements. It is to be hoped that an approach of the type given here may serve to establish the practical relevance of the Kronecker canonical form.

2. REGULAR SYSTEMS OF DIFFERENTIAL EQUATIONS

In this section we shall re-develop the Weierstrass theory for regular pencils $A-\lambda B$, the whole motivation being derived from consideration of the $n \times n$ system

$$B\dot{x} = Ax + f. \tag{2.1}$$

The $n \times n$ matrix pencil $A-\lambda B$ is defined to be regular if $\det(A-\lambda B) \not\equiv 0$. If we assume a knowledge of the Jordan canonical form, as we shall, the Weierstrass theory is almost trivial particularly if approached from the standpoint of the differential system, but it is an essential introduction to the Kronecker theory. The system

(2.1) may be transformed into an equivalent system by pre-multiplication with a non-singular P and a transformation

$$x = Qy \tag{2.2}$$

of the variables with a non-singular Q. The equivalent system is

$$PBQ\dot{y} = PAQy + Pf \tag{2.3}$$

or

$$\tilde{B}\dot{y} = \tilde{A}y + g. \tag{2.4}$$

The aim is to choose P and Q so that B and A are as simple as possible. Corresponding to the transformation of the differential system we have the transformation

$$P(A-\lambda B)Q = \tilde{A}-\lambda\tilde{B} \tag{2.5}$$

of the matrix pencil. Notice that P and Q are independent of λ; indeed λ does not occur in the differential equations. In (2.5) $A-\lambda B$ is said to be <u>strictly equivalent</u> to $\tilde{A}-\lambda\tilde{B}$, the term strictly implying that P and Q are independent of λ. The transformation to (2.4) will usually take place in a number of elementary steps in some of which selected variables will remain unchanged. The corresponding matrix involved in the change of variables in such a stage will be block diagonal, the diagonal block being the identity matrix for the variables which are unchanged. We proceed to the general case via three easy stages.

(i) <u>Non-singular B</u>

In this case the system (2.1) may be transformed to

$$\dot{x} = B^{-1}Ax + B^{-1}f. \tag{2.6}$$

If J is the J.c.f. of $B^{-1}A$ we have

$$X^{-1}B^{-1}AX = J \tag{2.7}$$

for some non-singular X. Hence writing

$$x = Xy \tag{2.8}$$

(2.6) becomes

$$X\dot{y} = B^{-1}AXy + B^{-1}f \tag{2.9}$$

or

$$\dot{y} = X^{-1}B^{-1}AXy + X^{-1}B^{-1}f = Jy + X^{-1}B^{-1}f. \tag{2.10}$$

The differential system is now in the standard form. If we write

$$Q = X, \qquad P = X^{-1}B^{-1} = Q^{-1}B^{-1} \tag{2.11}$$

then

$$\dot{y} = Jy + Pf \tag{2.12}$$

$$J = X^{-1}B^{-1}AX = PAQ \tag{2.13}$$

$$I = X^{-1}B^{-1}BX = PBQ. \tag{2.14}$$

Hence

$$P(A-\lambda B)Q = J - \lambda I \tag{2.15}$$

and we have the Weierstrass canonical reduction of the pencil $A-\lambda B$.

(ii) Singular B but non-singular A

(2.1) may now be transformed to

$$A^{-1}B\dot{x} = x + A^{-1}f \tag{2.16}$$

and if $\tilde{J}$ is the J.c.f. of $A^{-1}B$ there is a non-singular X such that

$$X^{-1}A^{-1}BX = \tilde{J} . \tag{2.17}$$

Hence writing

$$x = Xz \tag{2.18}$$

(2.16) becomes

$$A^{-1}BX\dot{z} = Xz + A^{-1}f \tag{2.19}$$

$$\tilde{J}\dot{z} = z + X^{-1}A^{-1}f . \tag{2.20}$$

Now by assumption B is singular and hence $A^{-1}B$ and $\tilde{J}$ are singular. In $\tilde{J}$ therefore at least one of the elementary Jordan blocks is nil-potent, ie is of the form $J_k(0)$. For simplicity of presentation we assume that n = 6 and

$$\tilde{J} = \begin{bmatrix} J_3(\mu_1) & \\ & J_3(0) \end{bmatrix} \tag{2.21}$$

where $\mu_1 \neq 0$. (In general $\tilde{J}$ will contain several blocks associated with non-zero μ_i and possibly several nil-potent blocks). Since $\tilde{J}$ is singular we cannot bring (2.20) to standard form by pre-multiplying by $\tilde{J}^{-1}$. However we can bring it close to standard form by pre-multiplying with a block diagonal matrix with blocks which are the reciprocals of those associated with the non-zero μ_i and equal to I in those blocks associated with the nil-potent matrices. Thus in our case we premultiply (2.20) with

$$Y = \begin{bmatrix} J_3^{-1}(\mu_1) & \\ & I \end{bmatrix} \tag{2.22}$$

to obtain

$$\begin{bmatrix} I & \\ & J_3(0) \end{bmatrix} \dot{z} = \begin{bmatrix} J_3^{-1}(\mu_1) & \\ & I \end{bmatrix} z + YX^{-1}A^{-1}f\ . \tag{2.23}$$

Now

$$J_3^{-1}(\mu_1) = \begin{bmatrix} \lambda_1 & -\lambda_1^2 & \lambda_1^3 \\ & \lambda_1 & -\lambda_1^2 \\ & & \lambda_1 \end{bmatrix} \text{ where } \lambda_1 = \mu_1^{-1} \tag{2.24}$$

from which the general form of the inverse of a block of any order associated with a non-zero μ_i is immediately apparent. Clearly $J_3^{-1}(\mu_1) - \lambda_1 I$ is of rank two; hence $J_3^{-1}(\mu_1)$ is similar to $J_3(\lambda_1)$ and there is a non-singular Z s.t.

$$Z^{-1}\ J_3^{-1}(\mu_1)Z = J_3(\lambda_1). \tag{2.25}$$

Defining W by

$$W = \begin{bmatrix} Z & \\ & I \end{bmatrix}, \tag{2.26}$$

and writing

$$z = Wy \tag{2.27}$$

$$\begin{bmatrix} I & \\ & J_3(0) \end{bmatrix} \dot{y} = \begin{bmatrix} J_3(\lambda_1) & \\ & I \end{bmatrix} y + W^{-1}YX^{-1}A^{-1}f. \qquad (2.28)$$

The equations are now as close to standard form as is attainable since, speaking of the general case, we cannot transform the nil-potent blocks into identity matrices. Hence if we write

$$W^{-1}YX^{-1}A^{-1} = P, \qquad XW = Q, \qquad (2.29)$$

then

$$PBQ = \begin{bmatrix} I & \\ & J_3(0) \end{bmatrix}, \; PAQ = \begin{bmatrix} J_3(\lambda_1) & \\ & I \end{bmatrix} . \qquad (2.30)$$

The general case is obvious from this. The nil-potent blocks will be retained in PBQ and PAQ will contain all the blocks associated with finite λ_i. From the generalizations of (2.30) we have the Weierstrass canonical form of the pencil $A-\lambda B$. The λ_i correspond to the finite eigenvalues of $Av = \lambda Bv$ and the nil-potent matrices give the 'infinite' eigenvalues and their structure (ie the zero eigenvalues of $\mu Av = Bv$).

Returning to the differential equations we may write (2.28) in the form

$$\begin{bmatrix} I & \\ & J_3(0) \end{bmatrix} \dot{y} = \begin{bmatrix} J_3(\lambda_1) & \\ & I \end{bmatrix} y + g \quad . \qquad (2.31)$$

The variables associated with the blocks separate and the equations associated with the λ_i are exactly as in the explicit case. For the nil-potent blocks we have typically

$$\left.\begin{aligned} \dot{y}_5 &= y_4 + g_4 \\ \dot{y}_6 &= y_5 + g_5 \\ 0 &= y_6 + g_6 \end{aligned}\right\} \qquad (2.32)$$

giving

$$y_6 = -g_6, \quad y_5 = -g_5-\dot{g}_6, \quad y_4 = -g_4-\dot{g}_5-\ddot{g}_6 \ . \qquad (2.33)$$

The associated variables are uniquely determined; we cannot prescribe their initial values arbitrarily. Other than differentiability the g_i are not required to satisfy any relations for a solution to exist.

(iii) Singular A and B

Since we are considering regular pencils $\det(A-\lambda B) \not\equiv 0$; $A-pB$ is singular for only a finite number ($<n$) of values of p. Let us choose one for which $A-pB$ is not singular and make the transformation

$$x = ze^{pt}. \qquad (2.34)$$

This will prove only to be a 'temporary' measure, but it will provide our motivation. The equations become

$$B\dot{z} = (A-pB)z + e^{-pt}f. \qquad (2.35)$$

and writing this as

$$B\dot{z} = \tilde{A}z + \tilde{f}, \text{ where } \tilde{A} = A-pB, \ \tilde{f} = e^{-pt}f \qquad (2.36)$$

we have a system in which B is singular but A is not, ie case (ii) above. We could forget about the original system and continue with (2.36). In order to have a direct tie up with the Weierstrass theory we shall not do this, but shall continue with (2.35). If $\tilde{J}$ denotes the J.c.f. of $(A-pB)^{-1}B$ there is a non-singular X such that

$$X^{-1}(A-pB)^{-1}BX = \tilde{J} \tag{2.37}$$

and making the transformation

$$z = Xw \tag{2.38}$$

and premultiplying by $(A-pB)^{-1}$

$$\tilde{J}\dot{w} = w + X^{-1}(A-pB)^{-1}e^{-pt}f. \tag{2.39}$$

If now we write

$$w = ue^{-pt} \tag{2.40}$$

the equation becomes

$$\tilde{J}\dot{u} = (I+p\tilde{J})u + X^{-1}(A-pB)^{-1}f\ . \tag{2.41}$$

We could have arrived at this system if we had written (2.1) in the form

$$B\dot{x} = [(A-pB) + pB]\ x + f, \tag{2.42}$$

had premultiplied by $(A-pB)^{-1}$ and then X^{-1}, and had made the change of variables

$$x = Xu. \tag{2.43}$$

The introduction of z has served only to provide motivation. Now since B is singular, the same is true of $(A-pB)^{-1}B$ and of $\tilde{J}$, and hence $\tilde{J}$ will contain at least one nil-potent block. As before, for simplicity, we shall assume that it is of order 6 and contains just one nil-potent block and one other associated with μ_1. Equation (2.41) is therefore of the form

$$\begin{bmatrix} J_3(\mu_1) & \\ & J_3(0) \end{bmatrix} \dot{u} = \begin{bmatrix} I+pJ_3(\mu_1) & \\ & I+pJ_3(0) \end{bmatrix} u + X^{-1}(A-pB)^{-1}f. \quad (2.44)$$

If we premultiply now by

$$Y = \begin{bmatrix} J_3^{-1}(\mu_1) & \\ & [I+pJ_3(0)]^{-1} \end{bmatrix} \quad (2.45)$$

the system becomes

$$\begin{bmatrix} I & \\ & L \end{bmatrix} \dot{u}' = \begin{bmatrix} M(\mu_1) & \\ & I \end{bmatrix} u + YX^{-1}(A-pB)^{-1}f, \quad (2.46)$$

where

$$L = [I+pJ_3(0)]^{-1}J_3(0), M(\mu_1) = J_3^{-1}(\mu_1)[I+pJ_3(\mu_1)]. \quad (2.47)$$

In the general case $\dot{u}$ will be associated with a block diagonal matrix which is the unit matrix except where J has nil-potent blocks, while u is associated with a block diagonal matrix which has I in the position of the nil-potent blocks of J and blocks of the form $M(\mu_1)$ elsewhere. The system (2.46) may be further reduced to one in which the matrices multiplying $\dot{u}$ and u consist of pure Jordan blocks, those associated with $\dot{u}$ being nil-potent. Since

$$L = \begin{bmatrix} 0 & 1 & -p \\ 0 & 0 & 1 \\ 0 & 0 & 0 \end{bmatrix} \text{ and } M(\mu_1) = \begin{bmatrix} \lambda_1 & -\mu_1^{-2} & \mu_1^{-3} \\ & \lambda_1 & -\mu_1^{-2} \\ & & \lambda_1 \end{bmatrix}, \quad (2.48)$$

where $\lambda_1 = (1+p\mu_1)/\mu_1$ both L and $M(\mu_1)-\lambda_1 I$ are of rank two and there are non-singular matrices U and V such that

$$U^{-1}LU = J_3(0), \; V^{-1}M(\mu_1)V = J_3(\lambda_1). \quad (2.49)$$

Hence premultiplying with the matrix W^{-1} where

$$W = \begin{bmatrix} V & \\ & U \end{bmatrix} \tag{2.50}$$

and making the transformation

$$u = Wy \tag{2.51}$$

$$\begin{bmatrix} I & \\ & J_3(0) \end{bmatrix} \dot{y} = \begin{bmatrix} J_3(\lambda_1) & \\ & I \end{bmatrix} y + g, \tag{2.52}$$

so that the final form is the same as that for case (ii). If we write

$$P = W^{-1}YX^{-1}(A-pB)^{-1}, \quad Q = XW \tag{2.53}$$

$$PAQ = \begin{bmatrix} J_3(\lambda_1) & \\ & I \end{bmatrix}, PBQ = \begin{bmatrix} I & \\ & J_3(0) \end{bmatrix} \tag{2.54}$$

giving the canonical form for the regular pencil $A-\lambda B$. In the general case PBQ may have a number of nil-potent blocks and PAQ will have a number of Jordan blocks associated with finite λ_i. These correspond to the 'infinite' eigenvalues and the finite eigenvalues of $Av = \lambda Bv$ respectively. Notice that although the μ_i were functions of p the λ_i are not; also since A is singular at least one of the λ_i is zero and hence PAQ will also contain at least one nil-potent block. This will be significant for the physical problem from which the differential equations were derived, but hardly for the canonical form of the regular pencil. Singularity of A is, as it were, an 'accidental' feature which in the regular case could be removed by the simple substitution $x = ze^{pt}$ with almost any value of p. We can summarise the results for a regular system of differential equations as follows.

If B is non-singular then a solution not only exists but one can prescribe the initial values of the variables arbitrarily. When B is singular, solutions exist but one cannot prescribe the initial values arbitrarily. We saw that some of the transformed variables are uniquely determined. This means that there must be an equivalent number of linear relations between the initial values of the x_i, the right-hand-sides and their derivatives. The number of such relations is equal to the sum of the orders of the nil-potent blocks in the canonical transform of B; in general it is greater than the nullity of B since this is merely equal to the number of such blocks. In the homogeneous case, f = 0, and the x_i satisfy a number of linear homogeneous relations.

3. SINGULAR SYSTEMS

When the matrices A and B are rectangular $m \times n$ matrices or are square $n \times n$ matrices with $\det(A-\lambda B) \equiv 0$, the pencil $A-\lambda B$ is said to be singular. It will be convenient to discuss systems of differential equations

$$B\dot{x} = Ax + f \tag{3.1}$$

in which B and A are not necessarily square, and we shall refer to them as singular systems if the pencil $A-\lambda B$ is singular. In the rectangular case we obviously cannot reduce $A-\lambda B$ to a strictly equivalent $\tilde{A}-\lambda\tilde{B}$ which is block diagonal with square blocks; however this is also true when A and B are square if the pencil is singular.

Kronecker showed that in the singular case there exist non-singular square matrices P and Q of orders m and n respectively such that PAQ and PBQ are block diagonal with identical block structures but in addition to the square blocks which we have discussed in the regular case some of the diagonal blocks are now rectangular and of dimensions

$t \times (t+1)$ and/or $(t+1) \times t$. If we think in terms of the pencil $A-\lambda B$ then in the pencil $P(A-\lambda B)Q$ there are a number of blocks of the form typified when $t = 3$ by

$$\begin{bmatrix} -\lambda & 1 & 0 & 0 \\ 0 & -\lambda & 1 & 0 \\ 0 & 0 & -\lambda & 1 \end{bmatrix} \text{ and } \begin{bmatrix} -\lambda & 0 & 0 \\ 1 & -\lambda & 0 \\ 0 & 1 & -\lambda \\ 0 & 0 & 1 \end{bmatrix} \tag{3.2}$$

respectively. The corresponding blocks in A and B in the two cases are

$$\begin{bmatrix} 0 & 1 & 0 & 0 \\ 0 & 0 & 1 & 0 \\ 0 & 0 & 0 & 1 \end{bmatrix} \text{ and } \begin{bmatrix} 1 & 0 & 0 & 0 \\ 0 & 1 & 0 & 0 \\ 0 & 0 & 1 & 0 \end{bmatrix} ;$$

$$\begin{bmatrix} 0 & 0 & 0 \\ 1 & 0 & 0 \\ 0 & 1 & 0 \\ 0 & 0 & 1 \end{bmatrix} \text{ and } \begin{bmatrix} 1 & 0 & 0 \\ 0 & 1 & 0 \\ 0 & 0 & 1 \\ 0 & 0 & 0 \end{bmatrix} \tag{3.3}$$

respectively. Kronecker denoted the first of the blocks in (3.2) by L_3 and the second by L_3^T. He used the notation ε to denote the number of rows in the first type of block and the notation η to denote the number of columns in the second type of block. He included the cases $\varepsilon = 0$ and $\eta = 0$. Thus if there were two L_ε blocks with $\varepsilon = 0$ and three L_η^T blocks with $\eta = 0$ and these were regarded as being in the top left-hand corner of the canonical forms, the next block in the canonical form would have its upper left-hand element in position (4,3).

We illustrate this with the canonical form for the matrix pencil $A-\lambda B$ below.

$$\left[\begin{array}{cc|ccc|c|ccc|cc|ccc}
0 & 0 & & & & & & & & & & & & \\
0 & 0 & & & & & & & & & & & & \\
0 & 0 & & & & & & & & & & & & \\
\hline
 & & -\lambda & 1 & 0 & & & & & & & & & \\
 & & 0 & -\lambda & 1 & & & & & & & & & \\
\hline
 & & & & & -\lambda & & & & & & & & \\
 & & & & & 1 & & & & & & & & \\
\hline
 & & & & & & 1 & -\lambda & 0 & & & & & \\
 & & & & & & 0 & 1 & -\lambda & & & & & \\
 & & & & & & 0 & 0 & 1 & & & & & \\
\hline
 & & & & & & & & & 3-\lambda & 1 & & & \\
 & & & & & & & & & 0 & 3-\lambda & & & \\
\hline
 & & & & & & & & & & & 4-\lambda & 1 & 0 \\
 & & & & & & & & & & & 0 & 4-\lambda & 1 \\
 & & & & & & & & & & & 0 & 0 & 4-\lambda
\end{array}\right] \qquad (3.4)$$

Starting from the upper left-hand corner this has L_0, L_0, L_0^T, L_0^T, L_0^T, L_2, L_1^T, $I_3-\lambda J_3(0)$, $J_2(3)-\lambda I_2$, $J_3(4)-\lambda I_3$. Apart from the ordering of the blocks along the diagonal this canonical form is unique. The last three blocks are such as we have had in regular pencils. Kronecker discussed the canonical form in connexion with the elementary divisors of the homogeneous pencil $\mu A-\lambda B$. There are no elementary divisors corresponding to the L_ε and the L_η^T. The elementary divisors for the above example are therefore

$$\mu^3,\ (3\mu-\lambda)^2,\ (4\mu-\lambda)^3\ . \qquad (3.5)$$

The first of these is referred to as an <u>infinite elementary divisor</u> (in general there may be several such) and corresponds in the usual case of square matrices to the infinite eigenvalues of $Av = \lambda Bv$; the other two correspond to the finite eigenvalues. (In general there may be several such and they may include zero eigenvalues). The dimensions ε_i and η_j associated with the L_{ε_i} and $L_{\eta_j}^T$ are referred to as the <u>minimal column indices</u> and <u>minimal row indices</u>

respectively. Thus in example (3.4) we have $\varepsilon_1 = \varepsilon_2 = 0$, $\eta_1 = \eta_2 = \eta_3 = 0$, $\varepsilon_3 = 2$, $\eta_4 = 1$. When A and B are square there must be the same number of ε_i as of η_j but the two sets of indices will not, in general, be identical. The corresponding part of the pencil may be called the singular part, and the remainder, the regular part.

When working with a differential system it is natural to think of the transformed A and B as separate entities. Corresponding to (3.4) the A and B parts are

$$
\left[\begin{array}{cc|ccc|c|ccc|cc|ccc}
0 & 0 & & & & & & & & & & & & \\
0 & 0 & & & & & & & & & & & & \\
0 & 0 & & & & & & & & & & & & \\
\hline
 & & 0 & 1 & 0 & & & & & & & & & \\
 & & 0 & 0 & 1 & & & & & & & & & \\
\hline
 & & & & & 0 & & & & & & & & \\
 & & & & & 1 & & & & & & & & \\
\hline
 & & & & & & 1 & 0 & 0 & & & & & \\
 & & & & & & 0 & 1 & 0 & & & & & \\
 & & & & & & 0 & 0 & 1 & & & & & \\
\hline
 & & & & & & & & & 3 & 1 & & & \\
 & & & & & & & & & 0 & 3 & & & \\
\hline
 & & & & & & & & & & & 4 & 1 & 0 \\
 & & & & & & & & & & & 0 & 4 & 1 \\
 & & & & & & & & & & & 0 & 0 & 4
\end{array}\right]
$$

and

$$
\left[\begin{array}{cc|ccc|c|ccc|cc|ccc}
0 & 0 & & & & & & & & & & & & \\
0 & 0 & & & & & & & & & & & & \\
0 & 0 & & & & & & & & & & & & \\
\hline
 & & 1 & 0 & 0 & & & & & & & & & \\
 & & 0 & 1 & 0 & & & & & & & & & \\
\hline
 & & & & & 1 & & & & & & & & \\
 & & & & & 0 & & & & & & & & \\
\hline
 & & & & & & 0 & 1 & 0 & & & & & \\
 & & & & & & 0 & 0 & 1 & & & & & \\
 & & & & & & 0 & 0 & 0 & & & & & \\
\hline
 & & & & & & & & & 1 & 0 & & & \\
 & & & & & & & & & 0 & 1 & & & \\
\hline
 & & & & & & & & & & & 1 & 0 & 0 \\
 & & & & & & & & & & & 0 & 1 & 0 \\
 & & & & & & & & & & & 0 & 0 & 1
\end{array}\right]
\qquad (3.6)
$$

The Kronecker canonical form of $A-\lambda B$ gives us full information about the solution of the corresponding system of differential equations as we now show.

Suppose the $m \times n$ system

$$B\dot{x} = Ax + f \qquad (3.7)$$

already has B and A in the Kronecker canonical form. The nature of the solution corresponding to the variables associated with the regular part has already been dealt with. We have only to deal with the variables associated with the singular part. For simplicity of notation we do this for the matrices A and B above.

The first three equations have no contribution from Bx or Ax. Hence for compatibility we must have $f_1 = f_2 = f_3 = 0$. (This corresponds to three homogeneous linear relations between elements of the original right-hand side before reduction to canonical form). Each zero η_i requires one such relation. On the other hand x_1 and x_2 do not occur in the equations and can be taken to be arbitrary functions. Again each zero ε_i gives one such arbitrary choice.

Turning now to the L_3 block we have

$$\left.\begin{aligned} \dot{x}_3 &= x_4 + f_4 \\ \dot{x}_4 &= x_5 + f_5 \end{aligned}\right] . \qquad (3.8)$$

x_5 may be chosen to be an arbitrary function and x_4 and x_3 are then determined by quadratures. Notice in particular that the initial values of x_1, x_2, x_3, x_4 and x_5 may be prescribed arbitrarily. We have an analogous result for an L_ε block of any size, including, as we have already noticed, for an L_o block.

Finally consider the L_1^T block. This gives

$$\left.\begin{aligned} \dot{x}_6 &= f_6 \\ x_6 &= f_7 \end{aligned}\right] . \qquad (3.9)$$

The variable x_6 is completely prescribed; we cannot assign its initial value arbitrarily. Further we must have

$$\dot{f}_7 = f_6 \tag{3.10}$$

ie there must be a linear relation between an element of the right-hand side and the derivative of another element. In general for an L_η^T all η of the corresponding variables are uniquely determined (we cannot specify their initial values independently) and a relation must exist between elements of the right-hand side and their derivatives up to order η for compatibility.

The essential features introduced by singularity may be summarised as follows.

Corresponding to each L_ε the general solution contains one arbitrary function. It places no restriction on initial conditions or on the f_i.

Corresponding to each L_η^T , η of the variables are completely determined and cannot be assigned arbitrary initial values; a compatibility relation is required between

$$f_i,\ Df_{i+1},\ D^2f_{i+2},\ \dots\ ,\ D^\eta f_{i+\eta}\ \left(D = \frac{d}{dt}\right)$$

where L_η^T runs from row i to row i + η.

Although the reduction of A-λB to the Kronecker canonical form gives very complete information about the solution of the related system of differential equations the classical method of establishing it does not appear to be of much practical value. (Methods used for establishing the Jordan canonical form suffer from the same fault and indeed the more modern the exposition, the less algorithmic it tends to be). When the pencil is singular then rank (A-λB) is less than max(m, n) and hence there is always either a right-hand null vector or a left-hand null vector (or both), though these vectors are in general functions of λ. Hence we have either

$$(A-\lambda B)x(\lambda) = 0 \text{ or } y^T(\lambda)(A-\lambda B) = 0 \qquad (3.11)$$

or both. Restricting ourselves to vectors with elements which are polynomials in λ we may seek the vectors $x(\lambda)$ and $y(\lambda)$ of lowest degree. These degrees give the smallest ε and η respectively. The existence of the K.c.f. may be established by such considerations but it is difficult to construct a practical algorithm on such lines. We observe though that

$$\begin{bmatrix} -\lambda & 1 & 0 \\ 0 & -\lambda & 1 \end{bmatrix} \begin{bmatrix} 1 \\ \lambda \\ \lambda^2 \end{bmatrix} = 0. \qquad (3.12)$$

and this is the right null vector of lowest degree. In general L_ε has the right-null vector $(1, \lambda, \ldots, \lambda^\varepsilon)$; it has no left-null vectors. For L_η^T the roles of rows and columns are interchanged.

It is obvious that considerations of rank are implicitly involved in determining the K.c.f. This would be true in any case since they are implicit in determining the J.c.f. and this has to be done for the regular part of the pencil. The singular part presents additional difficulties associated with rank. We have already seen in connexion with regular pencils for which $\det(A) = \det(B) = 0$ it is simpler to restrict oneself to consideration of the differential system and our insistence on obtaining the W.c.f. of $A-\lambda B$ brought substantial complications. In the rest of the paper we consider the general solution of a singular system of differential equations as a problem in its own right.

4. SINGULAR SYSTEMS OF DIFFERENTIAL EQUATIONS

From now on

$$B\dot{x} = Ax + f \tag{4.1}$$

will denote a singular system, ie the pencil $A-\lambda B$ will be singular and we shall assume that A and B are rectangular matrices. This is convenient since, even if we start with an $n \times n$ system, we shall derive reduced systems for which the matrices may be rectangular.

We shall require one simple theorem from linear algebra. If A is an $m \times n$ matrix of rank r then there are non-singular matrices P and Q of orders m and n such that

$$PAQ = \left[\begin{array}{c|c} I_r & 0 \\ \hline 0 & 0 \end{array}\right]. \tag{4.2}$$

We have placed I_r in the upper left-hand corner. It could be in any r rows and columns and for convenience in subsequent work we shall sometimes place it in one of the other corners. Problems of numerical stability are always involved in rank determination. If we wish to emphasize them we may use the singular value decomposition rather than the above and we have

$$UAV = \left[\begin{array}{c|c} \Sigma_r & 0 \\ \hline 0 & 0 \end{array}\right], \tag{4.3}$$

where U and V are now unitary and Σ_r is the diagonal matrix given by the non-zero singular values. This has the advantage of revealing near rank deficiency and indeed if one wishes to achieve the reduction (4.2) in a stable manner one would probably proceed via the singular value decomposition. Again Σ_r could be in any corner to suit convenience.

It will be convenient to rename the original system

$$B_1\dot{x}^{(1)} = A_1x^{(1)} + f^{(1)} \tag{4.4}$$

and to regard B_1 and A_1 as matrices of dimension $M_1 \times N_1$. The reduction takes place in a number of stages. At a typical stage we shall be left with

$$B_s\dot{x}^{(s)} = A_sx^{(s)} + f^{(s)} \tag{4.5}$$

where B_s and A_s are matrices of dimension $M_s \times N_s$. At this stage $N_1 - N_s$ of the variables will already have been determined and $x^{(s)}$ will be of dimension N_s. The elements of the right-hand side $f^{(s)}$ will consist of linear combinations of the elements of $f^{(1)}$ and their derivatives as will appear below.

If B_s is of row nullity n_s there will be non-singular matrices P_s and Q_s such that

$$P_sB_sQ_s = \left[\begin{array}{c|c} I & 0 \\ \hline 0 & 0 \end{array}\right]\begin{array}{l} \}M_s-n_s \\ \\ \}n_s \end{array}, \quad P_sA_sQ_s = \left[\begin{array}{c} C_s \\ \hline D_s \end{array}\right]\begin{array}{l} \}M_s-n_s \\ \\ \}n_s \end{array}. \tag{4.6}$$

If D_s is of rank r_s (NB. $r_s \leq n_s$) then there will be non-singular matrices R_s and S_s such that

$$R_sD_sS_s = \left[\begin{array}{c|c} 0 & I \\ \hline 0 & 0 \end{array}\right]\begin{array}{l} \}r_s \\ \\ \}n_s-r_s \end{array}. \tag{4.7}$$

We emphasize that $n_s - r_s$ may be zero. Writing

$$\tilde{R}_s = \left[\begin{array}{c|c} I & 0 \\ \hline 0 & R_s \end{array}\right], \quad M_{s+1} = M_s-n_s, \; N_{s+1} = N_s - r_s \tag{4.8}$$

$$\tilde{R}_s P_s B_s Q_s S_s = \left[\begin{array}{c|c} B_{s+1} & H_s \\ \hline 0 & 0 \\ \hline 0 & 0 \end{array}\right] \begin{array}{l} \}M_{s+1} \\ \}r_s \\ \}n_s - r_s \end{array} \tag{4.9}$$

$$\tilde{R}_s P_s A_s Q_s S_s = \left[\begin{array}{c|c} A_{s+1} & F_s \\ \hline 0 & I \\ \hline 0 & 0 \end{array}\right] \begin{array}{l} \}M_{s+1} \\ \}r_s \\ \}n_s - r_s \end{array} \tag{4.10}$$

$$x^{(s)} = Q_s S_s \left[\begin{array}{c} x^{(s+1)} \\ \hline y^{(s)} \end{array}\right] \begin{array}{l} \}N_{s+1} \\ \}r_s \end{array}, \tilde{R}_s P_s f^{(s)} = \left[\begin{array}{c} g^{(s)} \\ \hline h^{(s)} \\ \hline i^{(s)} \end{array}\right] \begin{array}{l} \}M_{s+1} \\ \}r_s \\ \}n_s - r_s \end{array} \tag{4.11}$$

the system (4.5) is equivalent to

$$\left[\begin{array}{c|c} B_{s+1} & H_s \\ \hline 0 & 0 \\ \hline 0 & 0 \end{array}\right] \left[\begin{array}{c} \dot{x}^{(s+1)} \\ \dot{y}^{(s)} \end{array}\right] = \left[\begin{array}{c|c} A_{s+1} & F_s \\ \hline 0 & I \\ \hline 0 & 0 \end{array}\right] \left[\begin{array}{c} x^{(s+1)} \\ y^{(s)} \end{array}\right] + \left[\begin{array}{c} g^{(s)} \\ \hline h^{(s)} \\ \hline i^{(s)} \end{array}\right]. \tag{4.12}$$

From the last $n_s - r_s$ equations we have

$$i^{(s)} = 0 \tag{4.13}$$

ie $n_s - r_s$ linear relations to be satisfied between elements of $f^{(s)}$ for compatibility. From the next r_s equations

$$y^{(s)} = h^{(s)} \tag{4.14}$$

ie r_s of the current variables are fully determined in terms of linear combinations of the elements of $f^{(s)}$. Finally

$$\begin{aligned} B_{s+1}\dot{x}^{(s+1)} &= A_{s+1}x^{(s+1)} + g^{(s)} - H_s\dot{y}^{(s)} + F_s y^{(s)} \\ &= A_{s+1}x^{(s+1)} + f^{(s+1)} \end{aligned} \tag{4.15}$$

Notice that $f^{(s+1)}$ is derived from linear combinations of elements of $f^{(s)}$ and of their first derivatives (via the term $H_s\dot{y}^{(s)}$). This means that, in general, $f^{(s+1)}$ involves elements of $f^{(1)}$ and of their derivatives up to the sth. If $n_s = r_s$ no compatibility conditions arise from this step while if $r_s = 0$ no variables $y^{(s)}$ are determined and $f^{(s+1)}$ will not involve derivatives of elements of $f^{(s)}$. However when $r_s = 0$, this is certainly the last step.

The reduction terminates with the system

$$B_t\dot{x}^{(t)} = A_t x^{(t)} + f^{(t)} \tag{4.16}$$

when B_t has full row rank ie $n_t = 0$. At this stage we have

$$P_tB_tQ_t = [\underbrace{I}_{M_t} \mid \underbrace{O}_{N_t-M_t}]\}M_t,\ P_tA_tQ_t = C_t = [\underbrace{U_t}_{M_t} \mid \underbrace{V_t}_{N_t-M_t}]\}M_t \tag{4.17}$$

and if we write

$$Q_t x^{(t)} = x^{(t+1)},\ P_t f^{(t)} = f^{(t+1)} \tag{4.18}$$

and partition $x^{(t+1)}$ conformally into $[u^{(t+1)} \mid v^{(t+1)}]$, (4.16) gives

$$\dot{u}^{(t+1)} = U_t u^{(t+1)} + f^{(t+1)} + V_t\ v^{(t+1)}. \tag{4.19}$$

The variables in $v^{(t+1)}$ may be taken to be arbitrary functions and having done this (4.19) becomes a standard explicit system of differential equations of order M_t and its general solution may be expressed in terms of the J.c.f. of U_t.

It is natural to ask whether the eigenvalues of U_t will give the finite elementary divisors associated with the K.c.f. of the original A and B.

That this is not so is immediately evident if we consider the system

$$\dot{x}_1 = ax_1 + bx_2 + cx_3 + f \tag{4.20}$$

for which B is already of full row rank! If we adopt the policy above then x_2 and x_3 can be chosen to be arbitrary functions and having done this

$$x_1 = \alpha e^{at} + e^{at} \int_0^t (bx_2+cx_3+f)e^{-a\tau}d\tau \; . \tag{4.21}$$

This makes it appear as though the solution is associated with an elementary divisor λ-a. However this is not so unless both b and c are zero. For suppose b is non-zero; then the transformation

$$\bar{x}_1 = x_1, \; \bar{x}_2 = ax_1 + bx_2 + cx_3, \; \bar{x}_3 = x_3 \tag{4.22}$$

is non-singular and the system is equivalent to

$$\dot{\bar{x}}_1 = \bar{x}_2 + f \tag{4.23}$$

with the general solution

$$\bar{x}_3 \text{ and } \bar{x}_2 \text{ arbitrary functions} \tag{4.24}$$

$$\bar{x}_1 = \alpha + \int_0^t (\bar{x}_2+f)d\tau \tag{4.25}$$

and the special role of a has disappeared. In fact if

$$A = [a,b,c] \; , \; B = [1,0,0] \tag{4.26}$$

the K.c.f. of A-λB is

$$[0 \,\vdots\, -\lambda \;\; 1]. \tag{4.27}$$

It has no regular part (and hence no elementary divisors) and two L_ε matrices with $\varepsilon = 0$ and $\varepsilon = 1$. However, one might argue that the solution (4.21) with x_2 and x_3 arbitrary is as acceptable as (4.25) with $\bar{x}_2$ and $\bar{x}_3$ arbitrary. When b=c=0 on the other hand the K.c.f. is

$$[0 \mid 0 \mid a-\lambda] \tag{4.28}$$

and we have two L_ε matrices with $\varepsilon_1=\varepsilon_2=0$ and a finite elementary divisor $a-\lambda$; now e^{at} really does play a special role in the solution. Before discussing further the general solution of the system (4.16) we relate the reduction to this form with the K.c.f.

5. THE KRONECKER MINIMAL ROW INDICES

The reduction of the system (4.1) to the system (4.16) with B_t of full row rank determines a certain number of variables directly in terms of linear combinations of the original f_i and of their derivatives; it also produces compatibility conditions. Since the algorithm is based directly on the determination of row nullities and the 'remaining matrix' B_t is of full row rank one feels intuitively that it must expose the minimal row indices and the infinite elementary divisors of the pencil $A-\lambda B$ and it is easy to show that this is true. We do this by considering a further reduction which is wholly analogous to that used by Golub and Wilkinson [3] to establish the J.c.f. The final forms of A and B produced by the algorithm of section 4 if the transformations are applied at each stage to the full matrix are given typically when t = 4 by

$$A:\quad \begin{array}{c|c|c|c|l} & r_3 & r_2 & r_1 & \\ \hline A_4 & A_{43} & A_{42} & A_{41} & \\ \hline & \begin{matrix} I \\ 0 \end{matrix} & A_{32} & A_{31} & \}n_3 \\ \hline & & \begin{matrix} I \\ 0 \end{matrix} & A_{21} & \}n_2 \\ \hline & & & \begin{matrix} I \\ 0 \end{matrix} & \}n_1 \end{array} \qquad B:\quad \begin{array}{c|c|c|c|l} & r_3 & r_2 & r_1 & \\ \hline B_4 & B_{43} & B_{42} & B_{41} & \\ \hline & 0 & B_{32} & B_{31} & \}n_3 \\ \hline & & 0 & B_{21} & \}n_2 \\ \hline & & & 0 & \}n_1 \end{array} \tag{5.1}$$

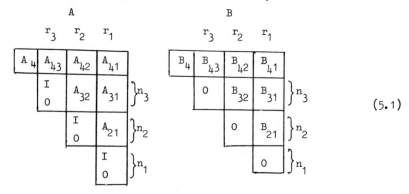

The matrices $B_{i+1,i}$ must be of full row rank otherwise n_i would not have been the full nullity at stage i. We have in fact

$$n_1 \geq r_1 \geq n_2 \geq r_2 \geq \ldots \geq n_t = 0 \quad . \tag{5.2}$$

The A_{ii} from their very form are of full column rank and B_t is, by definition, of full row rank.

The matrices $B_{t,t-1}$, $A_{t,t-1}$, $B_{t,t-2}$, $A_{t,t-2}$, ... B_{t1}, A_{t1} may now be annihilated successively. B_{ti} is annihilated by post- multiplications of A and B, which effectively subtract columns of B_t from those of B_{ti} and then A_{ti} is annihilated by pre- multiplications which effectively subtract rows of A_{ii} from those of A_{ti}. Blocks which are annihilated in any one step are not reintroduced in subsequent steps. The B matrix is now the direct sum of B_t and of the 'singular' part decomposed by the algorithm and the same applies to A.

The singular parts of A and B may now be further reduced to a form in which only the A_{ii} (which are obviously of full column rank) and the $B_{i+1,i}$ (which are of full row rank) remain. The blocks are annihilated in the order illustrated when t = 5 by A_{43}, B_{42}, A_{42}, B_{41}, A_{41}; A_{32}, B_{31}, A_{31}; A_{21}. In general A_{ij} is annihilated by pre-multiplications of A and B which effectively subtract rows of A_{ii} from rows of A_{ij} while B_{ij} is annihilated by post-multiplications of A and B which effectively subtract columns of $B_{i,i-1}$ from columns of B_{ij}. Notice that in both stages of this further reduction the A_{ii} and $B_{i+1,i}$ remain unaltered.

Finally the $B_{i+1,i}$ may be reduced to the form [I0] using both pre- multiplications and post-multiplications. Apart from the ordering of rows and columns the singular part is now in the K.c.f. as far as the minimal row indices and infinite elementary divisors are concerned. Since B_t is of full row rank the pencil can have no others. Reordering of the rows and columns shows that there are $n_i - r_i$ blocks L^T_{i-1} and $r_i - n_{i+1}$ infinite elementary divisors of degree i.

Two comments may be made about this further reduction.

(i) If we are primarily concerned with obtaining the solution of the differential system it serves little purpose. The form (5.1) is ideal as it stands and the minimal row indices are available without performing the further reduction.

(ii) The reduction of section 4 could be carried out using orthogonal transformations throughout if one had been content with diagonal matrices rather than the identity matrices in the A_{ii}. This is achieved by using the singular value decomposition at every step as in (4.3) rather than producing the form (4.2). For solving the differential system the diagonal form is almost as convenient and has the advantage of showing nearness to singularity. In the reduction of this section however orthogonal transformations have to be abandoned and one cannot achieve comparable numerical stability.

6. THE KRONECKER MINIMAL COLUMN INDICES

We have already shown in section 4 how to obtain a general solution of the remaining system (4.16) but this is not related to the K.c.f. and does not produce the Kronecker column indices or the finite elementary divisors.

Obviously, though, we can proceed with the system

$$B_t\dot{x}^{(t)} = A_t x^{(t)} + f^{(t)} \tag{6.1}$$

using an algorithm similar to that used for the minimal row indices. In fact if we reverse the roles of rows and columns and of A and B we shall obviously expose both the infinite elementary divisors of the pencil $B_t - \mu A_t$ (ie the 'zero elementary divisors' of $A_t - \lambda B_t$) and the minimal column indices. Naturally this algorithm can be used whether or not B_t is of full row rank. For this reason we describe it in general terms, and start again with the system

$$B_1\dot{x}^{(1)} = A_1x^{(1)} + f^{(1)} \tag{6.2}$$

where B_1 and A_1 are of order $M_1 \times N_1$ ignoring the fact that the B_1 and A_1 of (6.2) will usually be the B_t and A_t of (6.1) produced as the end product of our previous algorithm. For general organisational tidiness it is convenient now to work down from the top left-hand corner. However, even if we think in terms of solving the differential system rather than of deriving a canonical form we shall now be forced to retain the full $M_1 \times N_1$ system. In the earlier algorithm each stage determined certain variables (and compatibility conditions) and the values of the determined variables could be inserted into the remaining equations leaving us with a set of progressively lower order to deal with. The configuration at the beginning of stage s will be obvious if we display that at the beginning of stage 4 which is

$$\overset{B}{\left[\begin{array}{c|c|c|c} I0 & B_{12} & B_{13} & B_{14} \\ \hline 0 & I0 & B_{23} & B_{24} \\ \hline 0 & 0 & I0 & B_{34} \\ \hline 0 & 0 & 0 & B_4 \end{array}\right]}, \quad \overset{A}{\left[\begin{array}{c|c|c|c} 0 & A_{12} & A_{13} & A_{14} \\ \hline 0 & 0 & A_{23} & A_{24} \\ \hline 0 & 0 & 0 & A_{34} \\ \hline 0 & 0 & 0 & A_4 \end{array}\right]} . \tag{6.3}$$

The active part of the system is

$$B_s\dot{x}^{(s)} = A_sx^{(s)} + f^{(s)} \tag{6.4}$$

but any column transformations applied to A_s and B_s will have to be applied to the full M_1 rows.

If A_s is of column nullity n_s there will be non-singular matrices such that

$$P_sA_sQ_s = \left[\begin{array}{c|c} 0 & I \\ \hline 0 & 0 \end{array}\right], \quad P_sB_sQ_s = [\underbrace{D_s}_{n_s} \mid \underbrace{C_s}_{N_s-n_s}]. \tag{6.5}$$
$$\qquad\qquad \underbrace{\quad}_{n_s}\ \underbrace{\quad}_{N_s-n_s}$$

If D_s is of rank r_s (N.B. $r_s \leq n_s$) then there will be non-singular matrices R_s and S_s such that

$$R_sD_sS_s = \left[\begin{array}{c|c} I & 0 \\ \hline 0 & 0 \end{array}\right] . \tag{6.6}$$
$$\qquad\qquad \underbrace{\quad}_{r_s}\ \underbrace{\quad}_{n_s-r_s}$$

We have used a notation as close as possible to that in section 4 and it is now apparent that the matrices in (6.3) are indeed typical of the reduced A and B at the general stage. The process terminates when A_t is of full column rank i.e. $n_t = 0$; we have as before that

$$n_1 \geq r_1 \geq n_2 \geq r_2 \geq \dots \geq n_t = 0 \tag{6.7}$$

and that $A_{i,i+1}$ must be of full column rank at each stage otherwise n_i of the previous stage would not have been the full nullity of A_i. These comments apply whatever the nature of A_1 and B_1, but if A_1 and B_1 have in fact been the final matrices left by the previous algorithm, B_1 would be of full row rank and obviously the same must therefore be true of all the B_s produced by the new algorithm. On termination A_t is of full column rank and hence $M_t \geq N_t$; from our comments about the B_s we must have also have $M_t \leq N_t$. Hence on termination B_t and A_t are square and B_t is non-singular.

Accordingly

$$B_t \dot{x}^{(t)} = A_t x^{(t)} + f^{(t)} \tag{6.8}$$

is a square non-singular system and can be solved via the J.c.f. of $B_t^{-1}A_t$. The complete system can now be solved working from the bottom upwards. If we denote by $u^{(j)}$ the vector of values associated with the jth block column of the matrices of the form typified by (6.3) and by $g^{(i)}$ the vector of right-hand sides associated with the ith block row then

$$[I0]\dot{u}^{(i)} + \sum_{j=i+1}^{t-1} B_{ij}\dot{u}^{(j)} + B_{it}\dot{x}^{(t)} = \sum_{j=i+1}^{t-1} A_{ij}u^{(j)} + g^{(i)} \tag{6.9}$$

and in this equation all variables have been previously determined except those in $u^{(i)}$. Since the matrix multiplying $\dot{u}^{(i)}$ is [I0] only the first r_i components of $u^{(i)}$ are actually involved and the remaining $n_i - r_i$ components may be taken to be arbitrary functions. Each of the first r_i components of $u^{(i)}$ is then determined by a quadrature. Little convenience would be sacrificed if we were content to have diagonal matrices in place of the identity matrices in the transformed B and in this case we could use orthogonal transformations throughout.

If we are concerned only with the differential system the form (6.3) is perfectly adequate. However as with the earlier algorithm we could continue the reduction to the K.c.f. by eliminating first the coupling matrices A_{1t}, B_{1t}, ... , $A_{t-1,t}$, $B_{t-1,t}$ and then the $B_{ij}(j>i)$ and the $A_{ij}(j>i+1)$ and finally transforming the $A_{i,i+1}$ to the form

$$\begin{bmatrix} I \\ \\ 0 \end{bmatrix} \tag{6.10}$$

However the Kronecker column indices and the structure of the 'zero elementary divisors' of $A-\lambda B$ are known as soon as we have the n_i and r_i and there is no need to do this part of the reduction unless we really want the transformation matrices leading to the K.c.f. As with the earlier algorithms re-ordering of rows and columns shows that there are n_i-r_i blocks L_{i-1} and r_i-n_{i+1} zero elementary divisors of degree i.

The zero elementary divisors are not generally of any special significance as far as the theory of the differential system is concerned though they may be of importance in the underlying physical problem. However by reversing the roles of A and B as well as the roles of rows and columns our two algorithms become completely complementary and the zero elementary divisors are delivered automatically. Notice too that if we apply the second algorithm to $(A - kB)$ and B rather than to A and B it is the divisors of the form $(\lambda - k)^p$ which are exposed, i.e. those corresponding to the eigenvalues k in the case when A and B are square. This is analogous to the technique used by Golub and Wilkinson [3] to find the Jordan blocks associated with $(\lambda - k)$ for the standard eigenvalue problem $Ax = \lambda x$.

7. THE HOMOGENEOUS CASE

When f is zero the determination of the solution is simplified. At the sth stage of the first algorithm we are working with

$$B_s\dot{x}^{(s)} = Ax^{(s)} \qquad (7.1)$$

and when we perform the next reduction n_s-r_s of the equations ie those corresponding to (4.12) are null showing that the system (7.1) has (n_s-r_s)-fold linear dependence. The equations (4.14) become

$$y^{(s)} = 0 \qquad (7.2)$$

showing that the r_s variables in $y^{(s)}$ must all be zero. In the $n \times n$ singular case the original set of equations need not be linearly dependent but as we proceed at each stage we determine n_s transformed variables that are zero and when this is taken into account we must ultimately reach a system that is linearly dependent since otherwise $\det(A-\lambda B) \not\equiv 0$.

8. GENERAL COMMENTS

The part of the K.c.f. associated with the minimal indices may appear to be the exceptional part and the J.c.f. associated with the regular part as the heart of the K.c.f. It is salutory to realise that if we take an $m \times (m+1)$ pencil at random the probability is unity that the K.c.f. consists entirely of one L_m block and the corresponding remark applies to an $(m+1) \times m$ system. An $n \times n$ singular pencil must have at least one L_ε and one L_η^T.

A detailed discussion of the sensitivity of the K.c.f. is not the main object of this paper but the following observations give an indication of the considerations involved. If we take A and B to be both upper triangular with $a_{ss} = b_{ss} = 0$ and a_{ii} and b_{ii} otherwise non-zero then, in general, the K.c.f. will consist of an L_{s-1} and an L_{n-s}^T and there will be no regular part and no elementary divisors. However if we make perturbations in a_{ss} and b_{ss} however small then $A-\lambda B$ is regular and there are n finite linear elementary divisors. The pencil

$$\begin{bmatrix} 2 & 3 \\ 4 & 6 \end{bmatrix} - \lambda \begin{bmatrix} 1 & 1 \\ 2 & 2 \end{bmatrix} \tag{8.1}$$

has the K.c.f.

$$\begin{bmatrix} -\lambda & 1 \\ 0 & 0 \end{bmatrix} \tag{8.2}$$

consisting of an L_1 and an L_0^T. However random perturbation in A and B lead to a regular pencil, but with eigenvalues which are completely dependent on the perturbations. On the other hand the pencil

$$\begin{bmatrix} 2 & 4 \\ 6 & 12 \end{bmatrix} - \lambda \begin{bmatrix} 1 & 2 \\ 3 & 6 \end{bmatrix} \tag{8.3}$$

has the K.c.f.

$$\begin{bmatrix} 2-\lambda & 0 \\ 0 & 0 \end{bmatrix} \tag{8.4}$$

with an L_0 and an L_0^T and an elementary divisor $2-\lambda$. Small perturbations at random give a regular pencil but one of the elementary divisors is close to $2-\lambda$.

ACKNOWLEDGEMENTS

It is a pleasure to acknowledge the influence of P van Dooren on the final form taken by this paper. In [1] he gives an elegant derivation of the Kronecker canonical form based on the work of Kublanovskaya [5] and of Golub and Wilkinson [3].

REFERENCES

1. P. van Dooren (1977), The computation of Kronecker's canonical form of a singular period, Report TW34, Applied Mathematics and Programing Division, Katholicke Universiteit Leuven.

2. F. R Gantmacher (1959), Theory of matrices, Vol.II, Chelsea, New York.

3. G. H. Golub and J. H. Wilkinson (1976), Ill-conditioned eigensystems and the computation of the Jordan canonical form, SIAM Rev., 18, pp. 578-619.

4. L. Kronecker (1890), Algebraische Reduction der Schooren bilinearer Formen, S.-B. Akad. Berlin, pp. 763-776.

5. V. N. Kublanovskaya (1968), On a method of solving the complete eigenvalue problem for a degenerate matrix, USSR Comp. Math. and Math. Phys. 6, pp.1-14.

6. C. B. Moler and G. W. Stewart (1973), An algorithm for the generalized matrix eigenvalue problem $Ax = \lambda Bx$ SIAM J. Num. Anal., 10, pp. 241-256.

7. G. W. Stewart (1973), Introduction to Matrix Computations, Academic Press, New York and London.

8. J. H. Wilkinson (1965), The Algebraic Eigenvalue Problem, Oxford University Press.

9. ________________ and C. Reinsch (1971), Handbook for Automatic Computation, Vol.II: Linear Algebra, Springer-Verlag, Berlin and New York.

The author was supported by National Aeronautics and Space Administration Grant NSG 1443.

Department of Computer Science
Stanford University, Stanford,
California 94305.

Index